Science and Fiction

Series Editors

Mark Alpert, New York, USA

Philip Ball, London, UK

Gregory Benford, Irvine, USA

Michael Brotherton, Laramie, USA

Victor Callaghan, Colchester, UK

Amnon H. Eden, London, UK

Nick Kanas, Kentfield, USA

Rudy Rucker, Los Gatos, USA

Dirk Schulze-Makuch, Berlin, Germany

Rüdiger Vaas, Bietigheim-Bissingen, Germany

Ulrich Walter, Garching, Germany

Stephen Webb, Portsmouth, UK

Science and Fiction – A Springer Series

This collection of entertaining and thought-provoking books will appeal equally to science buffs, scientists and science-fiction fans. It was born out of the recognition that scientific discovery and the creation of plausible fictional scenarios are often two sides of the same coin. Each relies on an understanding of the way the world works, coupled with the imaginative ability to invent new or alternative explanations—and even other worlds. Authored by practicing scientists as well as writers of hard science fiction, these books explore and exploit the borderlands between accepted science and its fictional counterpart. Uncovering mutual influences, promoting fruitful interaction, narrating and analyzing fictional scenarios, together they serve as a reaction vessel for inspired new ideas in science, technology, and beyond.

Whether fiction, fact, or forever undecidable: the Springer Series "Science and Fiction" intends to go where no one has gone before!

Its largely non-technical books take several different approaches. Journey with their authors as they

- Indulge in science speculation – describing intriguing, plausible yet unproven ideas;
- Exploit science fiction for educational purposes and as a means of promoting critical thinking;
- Explore the interplay of science and science fiction – throughout the history of the genre and looking ahead;
- Delve into related topics including, but not limited to: science as a creative process, the limits of science, interplay of literature and knowledge;

Readers can look forward to a broad range of topics, as intriguing as they are important. Here just a few by way of illustration:

- Time travel, superluminal travel, wormholes, teleportation
- Extraterrestrial intelligence and alien civilizations
- Artificial intelligence, planetary brains, the universe as a computer, virtual worlds
- Non-anthropocentric viewpoints
- Synthetic biology, genetic engineering, developing nanotechnologies
- Eco/infrastructure/meteorite-impact disaster scenarios
- Future scenarios, transhumanism, posthumanism, intelligence explosion
- Consciousness and mind manipulation

Stephen Webb

Visions of Tomorrow

Exploring Classic Sci-Fi Stories Through the Lens of Modern Science

Stephen Webb
Lee on Solent, UK

ISSN 2197-1188   ISSN 2197-1196 (electronic)
Science and Fiction
ISBN 978-3-031-77599-4   ISBN 978-3-031-77600-7 (eBook)
https://doi.org/10.1007/978-3-031-77600-7

© The Editor(s) (if applicable) and The Author(s), under exclusive license to Springer Nature Switzerland AG 2024
This work is subject to copyright. All rights are solely and exclusively licensed by the Publisher, whether the whole or part of the material is concerned, specifically the rights of translation, reprinting, reuse of illustrations, recitation, broadcasting, reproduction on microfilms or in any other physical way, and transmission or information storage and retrieval, electronic adaptation, computer software, or by similar or dissimilar methodology now known or hereafter developed.
The use of general descriptive names, registered names, trademarks, service marks, etc. in this publication does not imply, even in the absence of a specific statement, that such names are exempt from the relevant protective laws and regulations and therefore free for general use.
The publisher, the authors and the editors are safe to assume that the advice and information in this book are believed to be true and accurate at the date of publication. Neither the publisher nor the authors or the editors give a warranty, expressed or implied, with respect to the material contained herein or for any errors or omissions that may have been made. The publisher remains neutral with regard to jurisdictional claims in published maps and institutional affiliations.

This Springer imprint is published by the registered company Springer Nature Switzerland AG
The registered company address is: Gewerbestrasse 11, 6330 Cham, Switzerland

If disposing of this product, please recycle the paper.

To the Ravenswood Road Groupies, especially Jenny, Nick, and Gurchetan—it was lovely to meet up after all those years

Preface

If you spend any time taking in the news—whether from old-school platforms such as radio and television or from those new-fangled social media outlets—you might suppose our current era differs from eras past. Pundits tell us we live in a more uncertain age. A more inventive age. More precarious. More savvy. More culturally vibrant. More … everything. A delusion, of course. We suffer from a form of recency bias. My anthology *New Light Through Old Windows*, which included a dozen classic science fiction tales from masters such as Arthur Conan Doyle and H. G. Wells, gave the lie to this delusion. Those stories, the earliest of which appeared in 1817, the latest in 1934, showed earlier authors grappling with questions that concern contemporary writers of speculative fiction. How can we prepare for the threat of an impact event or a pandemic? Where might technology lead us? What does it mean to be human? Those stories supplied answers that displayed no less insight than you will find in the pages of today's science fiction magazines.

In two key respects, though, our age *does* differ from those that have gone before. First, we have a deeper understanding of science. Second, we have better technology. The two factors intertwine: better technology enables us to do better science, and advances in science enable us to improve our technology. Each generation starts afresh on questions of ethics and morality; with science and technology each generation builds on and extends earlier work. We can identify progress.

My motivation for collecting the stories in *New Light Through Old Windows* stemmed from my fascination with the way earlier minds thought about the future, given the confines of their scientific and technical knowledge, but I also had a practical point in mind: the views of those authors give us a prism through which we can examine questions of current concern—topics such as artificial intelligence, big data, and robotics—given our own level of understanding (which, of course, will change and deepen over time). Examining old stories helps us refine our own thinking about the future.

Considerations of length forced me to omit from *New Light* authors I had hoped to include—notable names such as Chesterton, Poe, and Saki as well as lesser known writers such as Adeler, Breuer, and Waterloo. The present anthology allows me to make amends. The twelve stories here all come from the period 1845–1930 and from authors who did not appear in my earlier anthology.

My choice of stories might surprise readers who define "science fiction" in a narrow sense. Only Breuer's "The gostak and the doshes" appeared in an SF magazine (it first saw light in a 1930 issue of *Amazing Stories*). None of the other stories here appeared with the label "science fiction" attached to them. I contend, though, that the genre's boundaries have expanded since *Amazing Stories* launched in April 1926 and became the first magazine devoted to science fiction. All of the tales here (if suitably updated) would find a home in today's SF marketplace.

For example, Edward Bellamy's "With the eyes shut" discusses the societal implications of a specific invention; Edward Sabin's "The supersensitive golf ball" and Max Adeler's "The fortunate island", by way of contrast, tell of characters who find themselves in fantastical situations. We encounter both types of approach in modern science fiction. We see hard-edged technofuturism, of the type Bellamy would have approved, alongside gentler fantasies.

Or consider Stanley Waterloo's "Love and a triangle". The story discusses a question SF fans still ponder: how might we communicate with extraterrestrial creatures? Stories such as Algernon Blackwood's "A victim of higher space"; J. Arbuthnot Wilson's "Pausodyne: a great chemical discovery"; and Edgar Wallace's "The black grippe" each involve a conventional science fiction theme (higher dimensions, suspended animation, and pandemics, respectively). Stories of pure horror, such as Guy de Maupassant's "The Horla" and Edgar Allan Poe's "The facts in the case of M. Valdemar," appear with regularity in SF magazines. Even in mainstream works such as G. K. Chesterton's "The tremendous adventures of Major Brown" and Saki's "Filboid Studge, the story of a mouse that helped" we can discern elements of science fiction.

So I have no misgivings about appropriating these tales for a science fiction anthology. Indeed, despite whatever label literary critics might attach to these stories, many of them have influenced the field in some way.

Whether the stories involve hard prediction or light fantasy, horror or comedy, and regardless of the authors' intentions when writing them, these twelve tales all have one thing in common. One can see how, from our current vantage point, they shed light on the science and technology that will impact our lives in years to come. These 12 stories, then, all depict past visions of a possible future. Alongside each I present a brief commentary giving a modern take on the science or technology in the story. The commentaries cover three key themes implicit in the tales: physics and astronomy, computing, and biology.

Some stories stand up well, considering the changing stylistic tastes and social attitudes of the past 100–150 years. Others have aged with less grace, but deserve a reading because they pioneered some aspect of science fiction. The stories differ, then, in tone and style and technique, but I hope you will find at least one of these old tales to match your taste—and, despite their age, perhaps spark in you some new ways of thinking about the future.

Lee on Solent, UK   Stephen Webb

Contents

1	**Other Dimensions of Space**	1
	A Victim of Higher Space (Algernon Blackwood)	1
	Commentary	17
	Notes and Further Reading	21
	References	21
2	**Relativity**	23
	The Gostak and the Doshes (Miles J. Breuer)	23
	Commentary	38
	Notes and Further Reading	42
	References	42
3	**Messaging Extraterrestrial Intelligence**	43
	Love and a Triangle (Stanley Waterloo)	43
	Commentary	52
	Notes and Further Reading	57
	References	58
4	**First Contact**	59
	The Fortunate Island (Max Adeler)	59
	Chapter I. The Island	59
	Chapter II. The Castle of Baron Bors	78
	Chapter III. The Rescue	90
	Chapter IV. How the Professor Went Home	101
	Commentary	107
	Notes and Further Reading	110
	References	111

xii Contents

5 Applications of Machine Learning — 113
The Supersensitive Golf Ball (Edwin L. Sabin) — 113
Commentary — 118
Notes and Further Reading — 121
References — 122

6 The Science of Persuasion — 125
Filboid Studge, the Story of a Mouse that Helped (Saki) — 125
Commentary — 128
Notes and Further Reading — 131
References — 132

7 The Future of Science Publishing — 133
With the Eyes Shut (Edward Bellamy) — 133
Commentary — 148
Notes and Further Reading — 152
References — 153

8 Reality: Augmented, Virtual, and Mixed — 155
The Tremendous Adventures of Major Brown
(G. K. Chesterton) — 155
Commentary — 173
Notes and Further Reading — 177
References — 178

9 Pandemic — 179
The Black Grippe (Edgar Wallace) — 179
Commentary — 193
Notes and Further Reading — 197
References — 198

10 Suspended Animation — 199
Pausodyne: A Great Chemical Discovery
(J. Arbuthnot Wilson) — 199
Commentary — 213
Notes and Further Reading — 218
References — 218

11 Advances in Medical Technology — 221
The Facts in the Case of M. Valdemar (Edgar Allan Poe) — 221
Commentary — 229
Notes and Further Reading — 233
References — 234

12	**Humans Supplanted**	235
	The Horla (Guy de Maupassant)	235
	Commentary	255
	Notes and Further Reading	257
	References	258

1

Other Dimensions of Space

A Victim of Higher Space (Algernon Blackwood)

"There's a hextraordinary gentleman to see you, sir," said the new man.

"Why 'extraordinary'?" asked Dr. Silence, drawing the tips of his thin fingers through his brown beard. His eyes twinkled pleasantly. "Why 'extraordinary', Barker?" he repeated encouragingly, noticing the perplexed expression in the man's eyes.

"He's so—so thin, sir. I could hardly see 'im at all—at first. He was inside the house before I could ask the name," he added, remembering strict orders.

"And who brought him here?"

"He come alone, sir, in a closed cab. He pushed by me before I could say a word—making no noise not what I could hear. He seemed to move so soft like—".

The man stopped short with obvious embarrassment, as though he had already said enough to jeopardise his new situation, but trying hard to show that he remembered the instructions and warnings he had received with regard to the admission of strangers not properly accredited.

"And where is the gentleman now?" asked Dr. Silence, turning away to conceal his amusement.

"I really couldn't exactly say, sir. I left him standing in the 'all—".

The doctor looked up sharply. "But why in the hall, Barker? Why not in the waiting-room?" He fixed his piercing though kindly eyes on the man's face. "Did he frighten you?" he asked quickly.

"I think he did, sir, if I may say so. I seemed to lose sight of him, as it were—" The man stammered, evidently convinced by now that he had earned his dismissal. "He come in so funny, just like a cold wind," he added boldly, setting his heels at attention and looking his master full in the face.

The doctor made an internal note of the man's halting description; he was pleased that the slight signs of psychic intuition which had induced him to engage Barker had not entirely failed at the first trial. Dr. Silence sought for this qualification in all his assistants, from secretary to serving man, and if it surrounded him with a somewhat singular crew, the drawbacks were more than compensated for on the whole by their occasional flashes of insight.

"So the gentleman made you feel queer, did he?"

"That was it, I think, sir," repeated the man stolidly.

"And he brings no kind of introduction to me—no letter or anything?" asked the doctor, with feigned surprise, as though he knew what was coming.

The man fumbled, both in mind and pockets, and finally produced an envelope.

"I beg pardon, sir," he said, greatly flustered; "the gentleman handed me this for you."

It was a note from a discerning friend, who had never yet sent him a case that was not vitally interesting from one point or another.

"Please see the bearer of this note," the brief message ran, "though I doubt if even you can do much to help him."

John Silence paused a moment, so as to gather from the mind of the writer all that lay behind the brief words of the letter. Then he looked up at his servant with a graver expression than he had yet worn.

"Go back and find this gentleman," he said, "and show him into the green study. Do not reply to his question, or speak more than actually necessary; but think kind, helpful, sympathetic thoughts as strongly as you can, Barker. You remember what I told you about the importance of thinking, when I engaged you. Put curiosity out of your mind, and think gently, sympathetically, affectionately, if you can."

He smiled, and Barker, who had recovered his composure in the doctor's presence, bowed silently and went out.

There were two different reception-rooms in Dr. Silence's house. One (intended for persons who imagined they needed spiritual assistance when really they were only candidates for the asylum) had padded walls, and was well supplied with various concealed contrivances by means of which sudden violence could be instantly met and overcome. It was, however, rarely used. The other, intended for the reception of genuine cases of spiritual distress and out-of-the-way afflictions of a psychic nature, was entirely draped and furnished in a soothing deep green, calculated to induce calmness and repose of mind. And this room was the one in which Dr. Silence interviewed the majority of his "queer" cases, and the one into which he had directed Barker to show his present caller.

To begin with, the armchair in which the patient was always directed to sit, was nailed to the floor, since its immovability tended to impart this same excellent characteristic to the occupant. Patients invariably grew excited when talking about themselves, and their excitement tended to confuse their thoughts and to exaggerate their language. The inflexibility of the chair helped to counteract this. After repeated endeavours to drag it forward, or push it back, they ended by resigning themselves to sitting quietly. And with the futility of fidgeting there followed a calmer state of mind.

Upon the floor, and at intervals in the wall immediately behind, were certain tiny green buttons, practically unnoticeable, which on being pressed permitted a soothing and persuasive narcotic to rise invisibly about the occupant of the chair. The effect upon the excitable patient was rapid, admirable, and harmless. The green study was further provided with a secret spy-hole; for John Silence liked when possible to observe his patient's face before it had assumed that mask the features of the human countenance invariably wear in the presence of another person. A man sitting alone wears a psychic expression; and this expression is the man himself. It disappears the moment another person joins him. And Dr. Silence often learned more from a few moments' secret observation of a face than from hours of conversation with its owner afterwards.

A very light, almost a dancing, step followed Barker's heavy tread towards the green room, and a moment afterwards the man came in and announced that the gentleman was waiting. He was still pale and his manner nervous.

"Never mind, Barker," the doctor said kindly; "if you were not psychic the man would have had no effect upon you at all. You only need training and development. And when you have learned to interpret these feelings and sensations better, you will feel no fear, but only a great sympathy."

"Yes, sir; thank you, sir!" And Barker bowed and made his escape, while Dr. Silence, an amused smile lurking about the corners of his mouth, made his way noiselessly down the passage and put his eye to the spy-hole in the door of the green study.

This spy-hole was so placed that it commanded a view of almost the entire room, and, looking through it, the doctor saw a hat, gloves, and umbrella lying on a chair by the table, but searched at first in vain for their owner.

The windows were both closed and a brisk fire burned in the grate. There were various signs—signs intelligible at least to a keenly intuitive soul—that the room was occupied, yet so far as human beings were concerned, it was empty, utterly empty. No one sat in the chairs; no one stood on the mat before the fire; there was no sign even that a patient was anywhere close against the wall, examining the Bocklin reproductions—as patients so often did when they thought they were alone—and therefore rather difficult to see from the spy-hole. Ordinarily speaking, there was no one in the room. It was undeniable.

Yet Dr. Silence was quite well aware that a human being was in the room. His psychic apparatus never failed in letting him know the proximity of an incarnate or discarnate being. Even in the dark he could tell that. And he now knew positively that his patient—the patient who had alarmed Barker, and had then tripped down the corridor with that dancing footstep—was somewhere concealed within the four walls commanded by his spy-hole. He also realised—and this was most unusual—that this individual whom he desired to watch knew that he was being watched. And, further, that the stranger himself was also watching! In fact, that it was he, the doctor, who was being observed—and by an observer as keen and trained as himself.

An inkling of the true state of the case began to dawn upon him, and he was on the verge of entering—indeed, his hand already touched the doorknob—when his eye, still glued to the spy-hole, detected a slight movement. Directly opposite, between him and the fireplace, something stirred. He watched very attentively and made certain that he was not mistaken. An object on the mantelpiece—it was a blue vase—disappeared from view. It passed out of sight together with the portion of the marble mantelpiece on which it rested. Next, that part of the fire and grate and brass fender immediately below it vanished entirely, as though a slice had been taken clean out of them.

Dr. Silence then understood that something between him and these objects was slowly coming into being, something that concealed them and obstructed his vision by inserting itself in the line of sight between them and himself.

He quietly awaited further results before going in.

First he saw a thin perpendicular line tracing itself from just above the height of the clock and continuing downwards till it reached the woolly firemat. This line grew wider, broadened, grew solid. It was no shadow; it was something substantial. It defined itself more and more. Then suddenly, at the top of the line, and about on a level with the face of the clock, he saw a round luminous disc gazing steadily at him. It was a human eye, looking straight into his own, pressed there against the spy-hole. And it was bright with intelligence. Dr. Silence held his breath for a moment—and stared back at it.

Then, like someone moving out of deep shadow into light, he saw the figure of a man come sliding sideways into view, a whitish face following the eye, and the perpendicular line he had first observed broadening out and developing into the complete figure of a human being. It was the patient. He had apparently been standing there in front of the fire all the time. A second eye had followed the first, and both of them stared steadily at the spy-hole, sharply concentrated, yet with a sly twinkle of humour and amusement that made it impossible for the doctor to maintain his position any longer.

He opened the door and went in quickly. As he did so he noticed for the first time the sound of a German band coming in gaily through the open ventilators. In some intuitive, unaccountable fashion the music connected itself with the patient he was about to interview. This sort of prevision was not unfamiliar to him. It always explained itself later.

The man, he saw, was of middle age and of very ordinary appearance; so ordinary, in fact, that he was difficult to describe—his only peculiarity being his extreme thinness. Pleasant—that is, good—vibrations issued from his atmosphere and met Dr. Silence as he advanced to greet him, yet vibrations alive with currents and discharges betraying the perturbed and disordered condition of his mind and brain. There was evidently something wholly out of the usual in the state of his thoughts. Yet, though strange, it was not altogether distressing; it was not the impression that the broken and violent atmosphere of the insane produces upon the mind. Dr. Silence realised in a flash that here was a case of absorbing interest that might require all his powers to handle properly.

"I was watching you through my little peep-hole—as you saw," he began, with a pleasant smile, advancing to shake hands. "I find it of the greatest assistance sometimes—".

But the patient interrupted him at once. His voice was hurried and had odd, shrill changes in it, breaking from high to low in unexpected fashion. One moment it thundered, the next it almost squeaked.

"I understand without explanation," he broke in rapidly. "You get the true note of a man in this way—when he thinks himself unobserved. I quite agree. Only, in my case, I fear, you saw very little. My case, as you of course grasp, Dr. Silence, is extremely peculiar, uncomfortably peculiar. Indeed, unless Sir William had positively assured me—".

"My friend has sent you to me," the doctor interrupted gravely, with a gentle note of authority, "and that is quite sufficient. Pray, be seated, Mr—".

"Mudge—Racine Mudge," returned the other.

"Take this comfortable one, Mr. Mudge," leading him to the fixed chair, "and tell me your condition in your own way and at your own pace. My whole day is at your service if you require it."

Mr. Mudge moved towards the chair in question and then hesitated.

"You will promise me not to use the narcotic buttons," he said, before sitting down. "I do not need them. Also I ought to mention that anything you think of vividly will reach my mind. That is apparently part of my peculiar case." He sat down with a sigh and arranged his thin legs and body into a position of comfort. Evidently he was very sensitive to the thoughts of others, for the picture of the green buttons had only entered the doctor's mind for a second, yet the other had instantly snapped it up. Dr. Silence noticed, too, that Mr. Mudge held on tightly with both hands to the arms of the chair.

"I'm rather glad the chair is nailed to the floor," he remarked, as he settled himself more comfortably. "It suits me admirably. The fact is—and this is my case in a nutshell—which is all that a doctor of your marvellous development requires—the fact is, Dr. Silence, I am a victim of Higher Space. That's what's the matter with me—Higher Space!"

The two looked at each other for a space in silence, the little patient holding tightly to the arms of the chair which "suited him admirably," and looking up with staring eyes, his atmosphere positively trembling with the waves of some unknown activity; while the doctor smiled kindly and sympathetically, and put his whole person as far as possible into the mental condition of the other.

"Higher Space," repeated Mr. Mudge, "that's what it is. Now, do you think you can help me with that?"

There was a pause during which the men's eyes steadily searched down below the surface of their respective personalities. Then Dr. Silence spoke.

"I am quite sure I can help," he answered quietly; "sympathy must always help, and suffering always owns my sympathy. I see you have suffered cruelly. You must tell me all about your case, and when I hear the gradual steps by which you reached this strange condition, I have no doubt I can be of assistance to you."

He drew a chair up beside his interlocutor and laid a hand on his shoulder for a moment. His whole being radiated kindness, intelligence, desire to help.

"For instance," he went on, "I feel sure it was the result of no mere chance that you became familiar with the terrors of what you term Higher Space; for Higher Space is no mere external measurement. It is, of course, a spiritual state, a spiritual condition, an inner development, and one that we must recognise as abnormal, since it is beyond the reach of the world at the present stage of evolution. Higher Space is a mythical state."

"Oh!" cried the other, rubbing his birdlike hands with pleasure, "the relief it is to be able to talk to some one who can understand! Of course what you say is the utter truth. And you are right that no mere chance led me to my present condition, but, on the other hand, prolonged and deliberate study. Yet chance in a sense now governs it. I mean, my entering the condition of Higher Space seems to depend upon the chance of this and that circumstance. For instance, the mere sound of that German band sent me off. Not that all music will do so, but certain sounds, certain vibrations, at once key me up to the requisite pitch, and off I go. Wagner's music always does it, and that band must have been playing a stray bit of Wagner. But I'll come to all that later. Only first, I must ask you to send away your man from the spy-hole."

John Silence looked up with a start, for Mr. Mudge's back was to the door, and there was no mirror. He saw the brown eye of Barker glued to the little circle of glass, and he crossed the room without a word and snapped down the black shutter provided for the purpose, and then heard Barker snuffle away along the passage.

"Now," continued the little man in the chair, "I can begin. You have managed to put me completely at my ease, and I feel I may tell you my whole case without shame or reserve. You will understand. But you must be patient with me if I go into details that are already familiar to you—details of Higher Space, I mean—and if I seem stupid when I have to describe things that transcend the power of language and are really therefore indescribable."

"My dear friend," put in the other calmly, "that goes without saying. To know Higher Space is an experience that defies description, and one is obliged to make use of more or less intelligible symbols. But, pray, proceed. Your vivid thoughts will tell me more than your halting words."

An immense sigh of relief proceeded from the little figure half lost in the depths of the chair. Such intelligent sympathy meeting him half-way was a new experience to him, and it touched his heart at once. He leaned back, relaxing his tight hold of the arms, and began in his thin, scale-like voice.

"My mother was a Frenchwoman, and my father an Essex bargeman," he said abruptly. "Hence my name—Racine and Mudge. My father died before I ever saw him. My mother inherited money from her Bordeaux relations, and when she died soon after, I was left alone with wealth and a strange freedom. I had no guardian, trustees, sisters, brothers, or any connection in the world to look after me. I grew up, therefore, utterly without education. This much was to my advantage; I learned none of that deceitful rubbish taught in schools, and so had nothing to unlearn when I awakened to my true love—mathematics, higher mathematics and higher geometry. These, however, I seemed to know instinctively. It was like the memory of what I had deeply studied before; the principles were in my blood, and I simply raced through the ordinary stages, and beyond, and then did the same with geometry. Afterwards, when I read the books on these subjects, I understood how swift and undeviating the knowledge had come back to me. It was simply memory. It was simply re-collecting the memories of what I had known before in a previous existence and required no books to teach me."

In his growing excitement, Mr. Mudge attempted to drag the chair forward a little nearer to his listener, and then smiled faintly as he resigned himself instantly again to its immovability, and plunged anew into the recital of his singular "disease".

"The audacious speculations of Bolyai, the amazing theories of Gauss—that through a point more than one line could be drawn parallel to a given line; the possibility that the angles of a triangle are together greater than two right angles, if drawn upon immense curvatures—the breathless intuitions of Beltrami and Lobatchewsky—all these I hurried through, and emerged, panting but unsatisfied, upon the verge of my—my new world, my Higher Space possibilities—in a word, my disease!

"How I got there," he resumed after a brief pause, during which he appeared to be listening intently for an approaching sound, "is more than I can put intelligibly into words. I can only hope to leave your mind with an intuitive comprehension of the possibility of what I say.

"Here, however, came a change. At this point I was no longer absorbing the fruits of studies I had made before; it was the beginning of new efforts to learn for the first time, and I had to go slowly and laboriously through terrible work. Here I sought for the theories and speculations of others. But books were few and far between, and with the exception of one man—a 'dreamer', the world called him—whose audacity and piercing intuition amazed and delighted me beyond description, I found no one to guide or help.

"You, of course, Dr. Silence, understand something of what I am driving at with these stammering words, though you cannot perhaps yet guess what depths of pain my new knowledge brought me to, nor why an acquaintance with a new development of space should prove a source of misery and terror."

Mr. Racine Mudge, remembering that the chair would not move, did the next best thing he could in his desire to draw nearer to the attentive man facing him, and sat forward upon the very edge of the cushions, crossing his legs and gesticulating with both hands as though he saw into this region of new space he was attempting to describe, and might any moment tumble into it bodily from the edge of the chair and disappear form view. John Silence, separated from him by three paces, sat with his eyes fixed upon the thin white face opposite, noting every word and every gesture with deep attention.

"This room we now sit in, Dr. Silence, has one side open to space—to Higher Space. A closed box only seems closed. There is a way in and out of a soap bubble without breaking the skin."

"You tell me no new thing," the doctor interposed gently.

"Hence, if Higher Space exists and our world borders upon it and lies partially in it, it follows necessarily that we see only portions of all objects. We never see their true and complete shape. We see their three measurements, but not their fourth. The new direction is concealed from us, and when I hold this book and move my hand all round it I have not really made a complete circuit. We only perceive those portions of any object which exist in our three dimensions; the rest escapes us. But, once we learn to see in Higher Space, objects will appear as they actually are. Only they will thus be hardly recognisable!

"Now, you may begin to grasp something of what I am coming to."

"I am beginning to understand something of what you must have suffered," observed the doctor soothingly, "for I have made similar experiments myself, and only stopped just in time—".

"You are the one man in all the world who can hear and understand, and sympathise," exclaimed Mr. Mudge, grasping his hand and holding it tightly while he spoke. The nailed chair prevented further excitability.

"Well," he resumed, after a moment's pause, "I procured the implements and the coloured blocks for practical experiment, and I followed the instructions carefully till I had arrived at a working conception of four-dimensional space. The tesseract, the figure whose boundaries are cubes, I knew by heart. That is to say, I knew it and saw it mentally, for my eye, of course, could never take in a new measurement, or my hands and feet handle it.

"So, at least, I thought," he added, making a wry face. "I had reached the stage, you see, when I could imagine in a new dimension. I was able to conceive the shape of that new figure which is intrinsically different to all we know—the shape of the tesseract. I could perceive in four dimensions. When, therefore, I looked at a cube I could see all its sides at once. Its top was not foreshortened, nor its farther side and base invisible. I saw the whole thing out flat, so to speak. And this tesseract was bounded by cubes! Moreover, I also saw its content—its insides."

"You were not yourself able to enter this new world," interrupted Dr. Silence.

"Not then. I was only able to conceive intuitively what it was like and how exactly it must look. Later, when I slipped in there and saw objects in their entirety, unlimited by the paucity of our poor three measurements, I very nearly lost my life. For, you see, space does not stop at a single new dimension, a fourth. It extends in all possible new ones, and we must conceive it as containing any number of new dimensions. In other words, there is no space at all, but only a spiritual condition. But, meanwhile, I had come to grasp the strange fact that the objects in our normal world appear to us only partially."

Mr. Mudge moved farther forward till he was balanced dangerously on the very edge of the chair. "From this starting point," he resumed, "I began my studies and experiments, and continued them for years. I had money, and I was without friends. I lived in solitude and experimented. My intellect, of course, had little part in the work, for intellectually it was all unthinkable. Never was the limitation of mere reason more plainly demonstrated. It was mystically, intuitively, spiritually that I began to advance. And what I learnt, and knew, and did is all impossible to put into language, since it all describes experiences transcending the experiences of men. It is only some of the results—what you would call the symptoms of my disease—that I can give you, and even these must often appear absurd contradictions and impossible paradoxes.

"I can only tell you, Dr. Silence"—his manner became exceedingly impressive—"that I reached sometimes a point of view whence all the great puzzle of the world became plain to me, and I understood what they call in the Yoga books 'The Great Heresy of Separateness'; why all great teachers have urged the necessity of man loving his neighbour as himself; how men are all really one; and why the utter loss of self is necessary to salvation and the discovery of the true life of the soul."

He paused a moment and drew breath.

"Your speculations have been my own long ago," the doctor said quietly. "I fully realise the force of your words. Men are doubtless not separate at all—in the sense they imagine—".

"All this about the very much Higher Space I only dimly, very dimly, conceived, of course," the other went on, raising his voice again by jerks; "but what did happen to me was the humbler accident of—the simpler disaster—oh, dear, how shall I put it—?"

He stammered and showed visible signs of distress.

"It was simply this," he resumed with a sudden rush of words, "that, accidentally, as the result of my years of experiment, I one day slipped bodily into the next world, the world of four dimensions, yet without knowing precisely how I got there, or how I could get back again. I discovered, that is, that my ordinary three-dimensional body was but an expression—a projection—of my higher four-dimensional body!

"Now you understand what I meant much earlier in our talk when I spoke of chance. I cannot control my entrance or exit. Certain people, certain human atmospheres, certain wandering forces, thoughts, desires even—the radiations of certain combinations of colour, and above all, the vibrations of certain kinds of music, will suddenly throw me into a state of what I can only describe as an intense and terrific inner vibration—and behold I am off! Off in the direction at right angles to all our known directions! Off in the direction the cube takes when it begins to trace the outlines of the new figure! Off into my breathless and semi-divine Higher Space! Off, inside myself, into the world of four dimensions!"

He gasped and dropped back into the depths of the immovable chair.

"And there," he whispered, his voice issuing from among the cushions, "there I have to stay until these vibrations subside, or until they do something which I cannot find words to describe properly or intelligibly to you—and then, behold, I am back again. First, that is, I disappear. Then I reappear."

"Just so," exclaimed Dr. Silence, "and that is why a few—".

"Why a few moments ago," interrupted Mr. Mudge, taking the words out of his mouth, "you found me gone, and then saw me return. The music of that wretched German band sent me off. Your intense thinking about me brought me back—when the band had stopped its Wagner. I saw you approach the peep-hole and I saw Barker's intention of doing so later. For me no interiors are hidden. I see inside. When in that state the content of your mind, as of your body, is open to me as the day. Oh, dear, oh, dear, oh, dear!"

Mr. Mudge stopped and again mopped his brow. A light trembling ran over the surface of his small body like wind over grass. He still held tightly to the arms of the chair.

"At first," he presently resumed, "my new experiences were so vividly interesting that I felt no alarm. There was no room for it. The alarm came a little later."

"Then you actually penetrated far enough into that state to experience yourself as a normal portion of it?" asked the doctor, leaning forward, deeply interested.

Mr. Mudge nodded a perspiring face in reply.

"I did," he whispered, "undoubtedly I did. I am coming to all that. It began first at night, when I realised that sleep brought no loss of consciousness—".

"The spirit, of course, can never sleep. Only the body becomes unconscious," interposed John Silence.

"Yes, we know that—theoretically. At night, of course, the spirit is active elsewhere, and we have no memory of where and how, simply because the brain stays behind and receives no record. But I found that, while remaining conscious, I also retained memory. I had attained to the state of continuous consciousness, for at night I regularly, with the first approaches of drowsiness, entered nolens volens the four-dimensional world.

"For a time this happened regularly, and I could not control it; though later I found a way to regulate it better. Apparently sleep is unnecessary in the higher—the four-dimensional—body. Yes, perhaps. But I should infinitely have preferred dull sleep to the knowledge. For, unable to control my movements, I wandered to and fro, attracted, owing to my partial development and premature arrival, to parts of this new world that alarmed me more and more. It was the awful waste and drift of a monstrous world, so utterly different to all we know and see that I cannot even hint at the nature of the sights and objects and beings in it. More than that, I cannot even remember them. I cannot now picture them to myself even, but can recall only the memory of the impression they made upon me, the horror and devastating terror of it all. To be in several places at once, for instance—".

"Perfectly," interrupted John Silence, noticing the increase of the other's excitement, "I understand exactly. But now, please, tell me a little more of this alarm you experienced, and how it affected you."

"It's not the disappearing and reappearing per se that I mind," continued Mr. Mudge, "so much as certain other things. It's seeing people and objects in their weird entirety, in their true and complete shapes, that is so distressing. It introduces me to a world of monsters. Horses, dogs, cats, all of which I loved; people, trees, children; all that I have considered beautiful in life—everything, from a human face to a cathedral—appear to me in a different shape and aspect to all I have known before. I cannot perhaps convince you why this should be terrible, but I assure you that it is so. To hear the human voice proceeding from this novel appearance which I scarcely recognise as a human body is ghastly, simply ghastly. To see inside everything and everybody is a form of insight peculiarly distressing. To be so confused in geography as to find myself one moment at the North Pole, and the next at Clapham Junction—or possibly at both places simultaneously—is absurdly terrifying. Your imagination will readily furnish other details without my multiplying my experiences now. But you have no idea what it all means, and how I suffer."

Mr. Mudge paused in his panting account and lay back in his chair. He still held tightly to the arms as though they could keep him in the world of sanity and three measurements, and only now and again released his left hand in order to mop his face. He looked very thin and white and oddly unsubstantial, and he stared about him as though he saw into this other space he had been talking about.

John Silence, too, felt warm. He had listened to every word and had made many notes. The presence of this man had an exhilarating effect upon him. It seemed as if Mr. Racine Mudge still carried about with him something of that breathless Higher-Space condition he had been describing. At any rate, Dr. Silence had himself advanced sufficiently far along the legitimate paths of spiritual and psychic transformations to realise that the visions of this extraordinary little person had a basis of truth for their origin.

After a pause that prolonged itself into minutes, he crossed the room and unlocked a drawer in a bookcase, taking out a small book with a red cover. It had a lock to it, and he produced a key out of his pocket and proceeded to open the covers. The bright eyes of Mr. Mudge never left him for a single second.

"It almost seems a pity," he said at length, "to cure you, Mr. Mudge. You are on the way to discovery of great things. Though you may lose your life in the process—that is, your life here in the world of three dimensions—you would lose thereby nothing of great value—you will pardon my apparent rudeness, I know—and you might gain what is infinitely greater. Your suffering, of course, lies in the fact that you alternate between the two worlds and are never wholly in one or the other. Also, I rather imagine, though I cannot be certain of this from any personal experiments, that you have here and there penetrated even into space of more than four dimensions, and have hence experienced the terror you speak of."

The perspiring son of the Essex bargeman and the woman of Normandy bent his head several times in assent, but uttered no word in reply.

"Some strange psychic predisposition, dating no doubt from one of your former lives, has favoured the development of your 'disease'; and the fact that you had no normal training at school or college, no leading by the poor intellect into the culs-de-sac falsely called knowledge, has further caused your exceedingly rapid movement along the lines of direct inner experience. None of the knowledge you have foreshadowed has come to you through the senses, of course."

Mr. Mudge, sitting in his immovable chair, began to tremble slightly. A wind again seemed to pass over his surface and again to set it curiously in motion like a field of grass.

"You are merely talking to gain time," he said hurriedly, in a shaking voice. "This thinking aloud delays us. I see ahead what you are coming to, only please be quick, for something is going to happen. A band is again coming down the street, and if it plays—if it plays Wagner—I shall be off in a twinkling."

"Precisely. I will be quick. I was leading up to the point of how to effect your cure. The way is this: You must simply learn to block the entrances."

"True, true, utterly true!" exclaimed the little man, dodging about nervously in the depths of the chair. "But how, in the name of space, is that to be done?"

"By concentration. They are all within you, these entrances, although outer cases such as colour, music and other things lead you towards them. These external things you cannot hope to destroy, but once the entrances are blocked, they will lead you only to bricked walls and closed channels. You will no longer be able to find the way."

"Quick, quick!" cried the bobbing figure in the chair. "How is this concentration to be effected?"

"This little book," continued Dr. Silence calmly, "will explain to you the way." He tapped the cover. "Let me now read out to you certain simple instructions, composed, as I see you divine, entirely from my own personal experiences in the same direction. Follow these instructions and you will no longer enter the state of Higher Space. The entrances will be blocked effectively."

Mr. Mudge sat bolt upright in his chair to listen, and John Silence cleared his throat and began to read slowly in a very distinct voice.

But before he had uttered a dozen words, something happened. A sound of street music entered the room through the open ventilators, for a band had begun to play in the stable mews at the back of the house—the March from Tannhäuser. Odd as it may seem that a German band should twice within the space of an hour enter the same mews and play Wagner, it was nevertheless the fact.

Mr. Racine Mudge heard it. He uttered a sharp, squeaking cry and twisted his arms with nervous energy round the chair. A piteous look that was not far from tears spread over his white face. Grey shadows followed it—the grey of fear. He began to struggle convulsively.

"Hold me fast! Catch me! For God's sake, keep me here! I'm on the rush already. Oh, it's frightful!" he cried in tones of anguish, his voice as thin as a reed.

Dr. Silence made a plunge forward to seize him, but in a flash, before he could cover the space between them, Mr. Racine Mudge, screaming and struggling, seemed to shoot past him into invisibility. He disappeared like an arrow from a bow propelled at infinite speed, and his voice no longer sounded in the external air, but seemed in some curious way to make itself heard somewhere within the depths of the doctor's own being. It was almost like a faint singing cry in his head, like a voice of dream, a voice of vision and unreality.

"Alcohol, alcohol!" it cried, "give me alcohol! It's the quickest way. Alcohol, before I'm out of reach!"

The doctor, accustomed to rapid decisions and even more rapid action, remembered that a brandy flask stood upon the mantelpiece, and in less than a second he had seized it and was holding it out towards the space above the chair recently occupied by the visible Mudge. Then, before his very eyes, and long ere he could unscrew the metal stopper, he saw the contents of the closed glass phial sink and lessen as though some one were drinking violently and greedily of the liquor within.

"Thanks! Enough! It deadens the vibrations!" cried the faint voice in his interior, as he withdrew the flask and set it back upon the mantelpiece. He understood that in Mudge's present condition one side of the flask was open to space and he could drink without removing the stopper. He could hardly have had a more interesting proof of what he had been hearing described at such length.

But the next moment—the very same moment it almost seemed—the German band stopped midway in its tune—and there was Mr. Mudge back in his chair again, gasping and panting!

"Quick!" he shrieked, "stop that band! Send it away! Catch hold of me! Block the entrances! Block the entrances! Give me the red book! Oh, oh, oh-h-h-h!!!"

The music had begun again. It was merely a temporary interruption. The Tannhäuser March started again, this time at a tremendous pace that made it sound like a rapid two-step as though the instruments played against time.

But the brief interruption gave Dr. Silence a moment in which to collect his scattering thoughts, and before the band had got through half a bar, he had flung forward upon the chair and held Mr. Racine Mudge, the struggling little victim of Higher Space, in a grip of iron. His arms went all round his diminutive person, taking in a good part of the chair at the same time. He was not a big man, yet he seemed to smother Mudge completely.

Yet, even as he did so, and felt the wriggling form underneath him, it began to melt and slip away like air or water. The wood of the armchair somehow disentangled itself from between his own arms and those of Mudge. The phenomenon known as the passage of matter through matter took place. The little man seemed actually to get mixed up in his own being. Dr. Silence could just see his face beneath him. It puckered and grew dark as though from some great internal effort. He heard the thin, reedy voice cry in his ear to "Block the entrances, block the entrances!" and then—but how in the world describe what is indescribable?

John Silence half rose up to watch. Racine Mudge, his face distorted beyond all recognition, was making a marvellous inward movement, as though doubling back upon himself. He turned funnel-wise like water in a whirling vortex, and then appeared to break up somewhat as a reflection breaks up and divides in a distorting convex mirror. He went neither forward nor backwards, neither to the right nor the left, neither up nor down. But he went. He went utterly. He simply flashed away out of sight like a vanishing projectile.

All but one leg! Dr. Silence just had the time and the presence of mind to seize upon the left ankle and boot as it disappeared, and to this he held on for several seconds like grim death. Yet all the time he knew it was a foolish and useless thing to do.

The foot was in his grasp one moment, and the next it seemed—this was the only way he could describe it—inside his own skin and bones, and at the same time outside his hand and all round it. It seemed mixed up in some amazing way with his own flesh and blood. Then it was gone, and he was tightly grasping a draught of heated air.

"Gone! gone! gone!" cried a thick, whispering voice, somewhere deep within his own consciousness. "Lost! lost! lost!" it repeated, growing fainter and fainter till at length it vanished into nothing and the last signs of Mr. Racine Mudge vanished with it.

John Silence locked his red book and replaced it in the cabinet, which he fastened with a click, and when Barker answered the bell he inquired if Mr. Mudge had left a card upon the table. It appeared that he had, and when the servant returned with it, Dr. Silence read the address and made a note of it. It was in North London.

"Mr. Mudge has gone," he said quietly to Barker, noticing his expression of alarm.

"He's not taken his 'at with him, sir."

"Mr. Mudge requires no hat where he is now," continued the doctor, stooping to poke the fire. "But he may return for it—".

"And the humbrella, sir."

"And the umbrella."

"He didn't go out my way, sir, if you please," stuttered the amazed servant, his curiosity overcoming his nervousness.

"Mr. Mudge has his own way of coming and going, and prefers it. If he returns by the door at any time remember to bring him instantly to me, and be kind and gentle with him and ask no questions. Also, remember, Barker, to think pleasantly, sympathetically, affectionately of him while he is away. Mr. Mudge is a very suffering gentleman."

Barker bowed and went out of the room backwards, gasping and feeling round the inside of his collar with three very hot fingers of one hand.

It was 2 days later when he brought in a telegram to the study. Dr. Silence opened it, and read as follows:

"Bombay. Just slipped out again. All safe. Have blocked entrances. Thousand thanks. Address Cooks, London.—MUDGE."

Dr. Silence looked up and saw Barker staring at him bewilderingly. It occurred to him that somehow he knew the contents of the telegram.

"Make a parcel of Mr. Mudge's things," he said briefly, "and address them Thomas Cook & Sons, Ludgate Circus. And send them there exactly a month from today and marked 'To be called for'."

"Yes, sir," said Barker, leaving the room with a deep sigh and a hurried glance at the waste-paper basket where his master had dropped the pink paper.

Algernon Henry Blackwood (1869–1951) was an English author who is now best remembered as a prolific writer of ghost stories and supernatural fiction. With tales such as "The Willows" he was an influence on authors such as H. P. Lovecraft and William Hope Hodgson. His stories inspired science fiction authors too, and in particular pulp writers such as Frank Belknap Long and Clark Ashton Smith.

Commentary

We live in a world of three space dimensions. The ancient Greeks knew this. Aristotle, for example, wrote the following. "A magnitude if divisible one way is a line, if two ways a surface, and if three a body. Beyond these there is no other magnitude, because the three dimensions are all that there are." For completeness, Aristotle might have added: "a magnitude that is not divisible is a point". And there we have it. A point has zero dimensions; a line has one dimension; a plane has two dimensions; and a solid body has three dimensions.

We live in three dimensions.

Despite that undeniable fact of our experience, mathematicians always wondered what a fourth spatial dimension—a direction perpendicular to the three familiar dimensions—might look like.

Edwin Abbott's 1884 satire *Flatland* gave the Victorians a taste of how to picture a fourth dimension. Imagine two-dimensional creatures living on a plane embedded in a higher, third dimension. Those creatures would have knowledge of the world in two dimensions. That third dimension? Unknowable to them. Picture a sphere hovering over their plane. The plane's inhabitants cannot see the sphere because it resides in that third dimension. For the two-dimensional creatures of flatland, the sphere is invisible. But if the sphere drops, and falls through the plane, it makes its presence known. When the sphere first touches the plane, a dot appears. As the sphere falls, inhabitants of the plane see a circular disk growing in size. The disk reaches its largest extent when the sphere is halfway through the plane. And as the sphere continues its fall, the disk shrinks to a point before disappearing altogether. That succession of cross sections—nothing; point; increasing disk; maximum disk; decreasing disk; point; nothing—can help two-dimensional creatures visualise a three-dimensional sphere.

As Blackwood's character Mr. Mudge states, a tesseract is a four-dimensional hypercube—the analogue of a square in two dimensions and a cube in three. We three-dimensional creatures can borrow the flatlander trick described above. We can visualise a tesseract by imagining it 'falling' through our space. (Try it!)

We can use another trick of visualisation. Take a point (0 dimensions) and sweep it through space to form a straight line segment (1 dimension). Take that line segment and sweep it perpendicular to itself through space to form a square (2 dimensions). Take that square and sweep it up through space to form a cube (3 dimensions). Now take that cube and sweep it through space in a perpendicular direction to form a tesseract (4 dimensions). See Fig. 1.1. The difficulty of this visualisation technique lies, of course, in that fourth step. We struggle because our three-dimensional world contains no direction perpendicular to the three sides of a cube and so our minds have no experience of the manipulations required.

Perhaps, then, given the difficulties, we should give up on thoughts of extra dimensions? Well, no. That would be premature.

Blackwood's fascination with extra dimensions stemmed in part because he believed they might play a role in the supernatural. In the 1920s, scientists began to wonder whether extra dimensions might play a *natural* rather than supernatural role. In 1921, the German mathematician Theodor Kaluza introduced an extra space dimension into general relativity. This allowed Kaluza to develop a theory that unified Einstein's theory of gravity with Maxwell's theory of electromagnetism. In Kaluza's approach you have a single theory in four space dimensions; then, when you drop back down to three dimension, you have separate theories of gravity and electromagnetism. What a terrific idea! You don't need to have a Nobel Prize in physics, though, to raise a serious objection: we don't see that fourth space dimension. In 1926, the

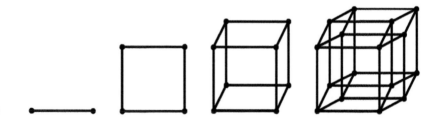

Fig. 1.1 If we sweep a point through space we generate a line segment. If we sweep the segment perpendicular to itself through space we generate a square. If we sweep the square through space perpendicular to its plane we generate a cube. If we could sweep the cube in a perpendicular direction we would generate a tesseract. (Own work)

Swedish physicist Oskar Klein came up with a plausible reason for why that extra dimension might hide from us. Klein argued that a fourth spatial dimension would be invisible if curled up into a circle with a small radius. In other words, if the fourth space dimension bent round on itself to become 'compact', as we now say, then we would not observe it. (See Fig. 1.2.)

Kaluza–Klein theory unified the two fundamental forces of nature—gravity and electromagnetism. Why, then, did their theory not become the standard theory of physics? Well, physicists discovered two *more* fundamental forces—the weak and strong forces. Traditional Kaluza–Klein theory said nothing about those other two forces and so their idea lost favour. Over time, though, physicists discovered that the same basic approach—formulating theories of physics in higher-dimensional spaces—offers a route to unifying all four fundamental forces.

In standard theories of particle physics, forces (such as the electromagnetic, weak or strong forces) act on the various charges possessed by particles of zero dimension (point particles such as electrons, quarks, neutrinos, and so on). In

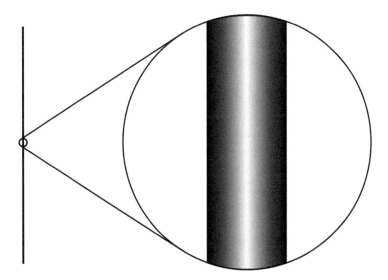

Fig. 1.2 From a distance, the structure on the left-hand side of the diagram appears to be a line: a one-dimensional object. When we look close, we see that the structure possesses a second dimension that has 'curled up' on itself, or compactified. The one-dimensional structure is in fact the two-dimensional object shown on the right-hand side of the diagram. It only appears one-dimensional when viewed from a distance. Perhaps the same is true of our universe. At each point in space there might exist an extra dimension (or indeed several extra dimensions). We would have to probe nature on small distance scales in order to observe those extra dimensions. (Credit: Alex Dunkel, CC BY-SA 4.0 International)

the 1980s, physicists began to go beyond those standard theories and instead construct so-called superstring theories. In these theories the basic building blocks of nature are not zero-dimensional point particles but rather one-dimensional strings. The different vibrations of these strings generate the particles we observe. One vibrational mode might give rise to a quark, another to an electron, yet another to a neutrino. And so on. There is a catch, though. For the mathematics to make sense, those strings must vibrate in higher-dimensional space. To be precise, six extra space dimensions must exist beyond the three we experience. If those extra dimensions compactify then that would account for their non-observation.

Since the 1980s, theoreticians have continued to explore fundamental physics in the setting of higher-dimensional spaces. For example, they have proposed the notion of 'branes'—spaces embedded within larger spaces. Think of Earth's surface. This is a two-dimensional space embedded in a larger three-dimensional space. One can think of Earth's surface here as a brane. (Of course, this simple analogy ignores the sophisticated mathematics required to understand branes fully.) By using the notions of strings and branes, theoreticians have begun to unravel some of the mysteries of black holes, for example, and these seemingly esoteric ideas have found fruitful applications in areas outside of particle physics.

But is there any experimental evidence that strings or branes or higher dimensions actually exist?

Well, no.

In general, these theories require that nature possesses a feature known as 'supersymmetry', or SUSY for short. Supersymmetry is the 'super' in 'superstrings'. SUSY demands that all the familiar point particles (electrons, quarks, neutrinos, and so on) must have supersymmetric partners (whose names have the letter 's' prefixed to the original particle: selectrons, squarks, sneutrinos, and so on). And all the familiar force carriers (photons, W particles, gluons, and so on) must have supersymmetric partners (whose names have the letters 'ino' postfixed to the names of the original force carrier: photinos, Winos, gluinos, and so on). Physicists hoped the Large Hadron Collider (LHC) would detect squarks or gluinos or some other particle in the menagerie, and thus provide evidence for supersymmetry. And if we discovered that nature displayed supersymmetry, then the notion of extra dimensions would become less outlandish. Some physicists even hoped the LHC would 'see' extra dimensions directly. But, as of this writing, particle physicists still desperately seek SUSY. And extra dimensions, if they exist, remain invisible.

Right now, the three dimensions of our familiar, everyday world comprise the entirety of the space dimensions we know for certain exist. The fourth dimension that Mr. Mudge could inhabit remains a world of imagination, not reality.

Notes and Further Reading

Aristotle, for example, wrote the following—The earliest discussion of the number of spatial dimensions goes back to the Greek philosopher Aristotle (384–322 BCE). Aristotle wrote about dimensions in Book I of *On the Heavens*. See, for example, the translation by Stocks (1922).

Edwin Abbott's 1884 satire—Edwin Abbott Abbott was an English schoolmaster. He wrote *Flatland: A Romance of Many Dimensions* (1884) under the pseudonym A. Square, and he used his novella to poke fun at Victorian culture. Nowadays we remember Abbott's work less for its satire than for its discussion of the dimensions of space and the possibility of higher dimensions.

Blackwood's fascination with extra dimensions—Blackwood was interested in the supernatural throughout his life, and to further this interest he joined The Ghost Club. (Other famous members of The Ghost Club included Sir William Crookes, Sir Oliver Lodge, and Sir Arthur Conan Doyle.) For a biography of Blackwood, see Ashley (2001).

Kaluza–Klein theory unified—For an accessible account of hidden dimensions, the Kaluza–Klein approach to unification, and of the ideas behind superstring theory, see Greene (1999).

Particle physicists are still desperately seeking SUSY—See Hershberger (2021) for a non-technical discussion of how physicists are searching for signs of supersymmetry in experiments taking place at CERN.

References

Abbott, A.A.: Flatland: A Romance of Many Dimensions. Seeley, London (1884)
Ashley, M.: Starlight Man: The Extraordinary Life of Algernon Blackwood. Constable & Robinson, London (2001)
Greene, B.R.: The Elegant Universe: Superstrings, Hidden Dimensions, and the Quest for the Ultimate Theory. W.W. Norton, New York (1999)
Hershberger, S.: The status of supersymmetry. Symmetry: Dimensions of Particle Physics. A joint Fermilab/SLAC publication. www.symmetrymagazine.org/article/the-status-of-supersymmetry (2021)
Stocks, J.L.: Aristotle's On the Heavens. Harvard University Press, Cambridge, MA. (1922) Freely available online from the Internet Archive

2

Relativity

The Gostak and the Doshes (Miles J. Breuer)

Let the reader suppose that somebody states "*The gostak distims the doshes*". You do not know what this means, nor do I. But if we assume that it is English, we know that the *doshes* are *distimmed* by the *gostak*. We know that one *distimmer* of the *doshes* is a *gostak*. If, moreover, doshes are galloons, we know that some galloons are distimmed by the gostak. And so we may go on, and so we often do go on.—Unknown writer quoted by Ogden and Richards, in *The Meaning of Meanings*, Harcourt Brace & Co., 1923; also by Walter N. Polakov in *Man and His Affairs*, Williams & Wilkins, 1925.

"Why! That is lifting yourself by your bootstraps!" I exclaimed in amazed incredulity. "It's absurd."

Woleshensky smiled indulgently. He towered in his chair as though in the infinite kindness of his vast mind there were room to understand and overlook all the foolish little foibles of all the weak little beings that called themselves men. A mathematical physicist lives in vast spaces where a light-year is a footstep, where universes are being born and blotted out, where space unrolls along a fourth dimension on a surface distended from a fifth. To him, human beings and their affairs do not loom very important.

"Relativity," he explained. In his voice there was a patient forbearance for my slowness of comprehension. "Merely relativity. It doesn't take much physical effort to make the moon move through the treetops, does it? Just enough to walk down the garden path."

I stared at him, and he continued: "If you had been born and raised on a moving train, no one could convince you that the landscape was not in rapid motion. Well, our conception of the universe is quite as relative as that. Sir Isaac Newton tried in his mathematics to express a universe as though beheld by an infinitely removed and perfectly fixed observer. Mathematicians since his time, realising the futility of such an effort, have taken into consideration that what things 'are' depends upon the person who is looking at them. They have tried to express common knowledge, such as the law of gravitation, in terms that would hold good to all observers. Yet their leader and culminating genius, Einstein, has been unable to express knowledge in terms of pure relativity; he has had to accept the velocity of light as an arbitrarily fixed constant. Why should the velocity of light be any more fixed and constant than any other quantity in the universe?"

"But, what's that got to do with going into the fourth dimension?" I broke in impatiently.

He continued as though I hadn't spoken.

"The thing that interests us now, and that mystifies modern mathematicians, is the question of movement, or more accurately, translation. Is there such a thing as absolute translation? Can there be movement—translation— except in relation to something else than the object that moves? All movement we know of is movement in relation to other objects, whether it be a walk down the street or the movement of the earth in its orbit around the sun. A change of relative position. But the mere translation of an isolated object existing alone in space is mathematically inconceivable, for there is no such thing as space in that sense."

"I thought you said something about going into another universe—" I interrupted again.

You can't argue with Woleshensky. His train of thought went on without a break.

"By translation we understand getting from one place to another. 'Going somewhere' originally meant a movement of our bodies. Yet, as a matter of fact, when we drive in an automobile we 'go somewhere' without moving our bodies at all. The scene is changed around us; we are somewhere else, and yet we haven't moved at all.

"Or suppose you could cast off gravitational attraction for a moment and let the earth rotate under you; you would be going somewhere and yet not moving—".

"But that is theory; you can't tinker with gravitation—".

"Every day you tinker with gravitation. When you start upward in an elevator, your pressure, not your weight, against the floor of it is increased;

apparent gravitation between you and the floor of the elevator is greater than before—and that's like gravitation is anyway: inertia and acceleration. But we are talking about translation. The position of everything in the universe must be referred to some sort of coordinates. Suppose we change the angle or direction of the coordinates: then you have 'gone somewhere' and yet you haven't moved, nor has anything else moved."

I looked at him, holding my head in my hands.

"I couldn't swear I understand that," I said slowly. "And I repeat, it looks like lifting yourself by your own bootstraps."

The homely simile did not dismay him. He pointed a finger at me as he spoke. "You've seen a chip of wood bobbing on the ripples of a pond. Now you think the chip is moving, now the water. Yet neither is moving; the only motion is of an abstract thing called a wave.

"You've seen those 'illusion' diagrams—for instance this one is a group of cubes. Make up your mind that you are looking down upon their upper surfaces and indeed they seem below you. Now change your mind and imagine that you are down below, looking up. Behold, you see their lower surfaces; you are indeed below them. You have 'gone somewhere', yet there has been no translation of anything. You have merely changed coordinates."

"Which do you think will drive me insane more quickly—if you show me what you mean, or if you keep on talking without showing me?"

"I'll try to show you. There are some types of mind, you know, that cannot grasp the idea of relativity. It isn't the mathematics involved that matters; it's just the inability of some types of mental organization to grasp the fact that the mind of the observer endows his environment with certain properties which have no absolute existence. Thus, when you walk through the garden at night, the moon floats from one treetop to another. Is your mind good enough to invert this: make the moon stand still and let the trees move backwards. Can you do that? If so, you can 'go somewhere' into another dimension."

Woleshensky rose and walked to the window. His office was an appropriate setting for such a modern discussion as ours—situated in a new, ultramodern building on the university campus, the varnish glossy, the walls clean, the books neatly arranged behind clean glass, the desk in the most orderly array; the office was just as precise and modern and wonderful as the mind of its occupant.

"When do you want to go?" he asked.

"Now!"

"Then, I have two more things to explain to you. The fourth dimension is just as much here as anywhere else. Right here around you and me things exist and go forward in the fourth dimension; but we do not see them and are not conscious of them because we are confined to our own three. Secondly, if we name the four coordinates as Einstein does, x, y, z, and t, then we exist in x, y, and z and move freely about in them, but are powerless to move in t. Why? Because t is the time dimension; and the time dimension is a difficult one for biological structures that depend on irreversible chemical reactions for their existence. But biochemical reactions can take place along any other dimensions as well as along t.

"Therefore, let us transform coordinates. Rotate the property of chemical irreversibility from t to z. Since we are organically able to exist (or at least to perceive) in only three dimensions at once, our new time dimension will be z. We shall be unconscious of z and cannot travel in it. Our activities and consciousness will occur along x, y, and t.

"According to fiction writers, to switch into the t dimension, some sort of apparatus with an electrical field ought to be necessary. It is not. You need nothing more to rotate into the t dimension than you do to stop the moon and make the trees move as you ride down the road; or than you do to turn the cubes upside down. It is a matter of relativity."

I had ceased trying to wonder or to understand.

"Show me!" was all I could gasp.

"The success of this experiment in changing from the z to the t coordinate has depended largely upon my lucky discovery of a favourable location. It is just as, when you want the moon to ride the tree tops successfully, there have to be favourable features in the topography or it won't work. The edge of this building and that little walk between the two rows of Norway poplars seems to be an angle between planes in the z and t dimensions. It seems to slope downward, does it not?—Now walk from here to the end and imagine yourself going upward. That is all. Instead of feeling this building behind and above you, conceive it as behind and below. Just as on your ride by moonlight, you must tell yourself that the moon is not moving while the trees ride by.— Can you do that? Go ahead then." He spoke confidently, as though he knew what would happen.

Half credulous, half wondering, I walked slowly out the door; I noticed that Woleshensky settled himself down at the table with a pad and a pencil to some kind of study, and forgot me before I had finished turning around. I looked curiously at the familiar wall of the building and the still more familiar poplar walk, expecting to see some strange scenery, some unknown view from another world. But there were the same old bricks and trees that I had known so long, though my disturbed and wondering frame of mind endowed them with a sudden strangeness and unwontedness. Things I had known for some years, they were, yet so powerfully had Woleshensky's arguments impressed me that I already fancied myself in a different universe. According to the conception of relativity, objects of the x, y, z universe ought to look different when viewed from the x, y, t universe.

Strange to say, I had no difficulty at all in imagining myself going upward on my stroll along the slope. I told myself that the building was behind and below me, and indeed it seemed real that it was that way. I walked some distance along the little avenue of poplars, which seemed familiar enough in all its details, though after a few minutes it struck me that the avenue seemed rather long. In fact, it was much longer than I had ever known it to be before.

With a queer Alice-in-Wonderland feeling I noted it stretching way on ahead of me. Then I looked back.

I gasped in astonishment. The building was indeed below me. I looked down upon it from the top of an elevation. The astonishment of that realization had barely broken over me when I admitted that there was a building down there; but what building? Not the new Morton Hall, at any rate. It was a long, three-story brick building, quite resembling Morton Hall, but it was not the same. And on beyond were trees with buildings among them; but it was not the campus that I knew.

I paused in a kind of panic. What was I to do now? Here I was in a strange place. How I had gotten there I had no idea. What ought I do about it? Where should I go? How was I to get back? Odd that I had neglected the precaution of how to get back. I surmised I was on the t dimension. Stupid blunder on my part, neglecting to find out how to get back.

I walked rapidly down the slope toward the building. Any hopes I might have had about its being Morton Hall were thoroughly dispelled in a moment. It was a totally strange building, old-fashioned looking. I had never seen it before in my life. Yet it looked ordinary and natural and was obviously a university classroom building.

I cannot tell whether it was an hour or a dozen that I spent walking frantically this way and that, trying to decide to go into this building or another, and at the last moment backing out in a sweat of hesitation. It seemed like a year but was probably only a few minutes. Then I noticed the people. They were mostly young people, of both sexes. Students, of course. Obviously, I was on a university campus. Perfectly natural, normal young people, they were. If I were really on the t dimension, it certainly resembled the z dimension very closely.

Finally I came to a decision. I could stand this no longer. I selected a solitary, quiet-looking man and stopped him.

"Where am I?" I asked.

He looked at me in astonishment. I waited for a reply, but he continued to gaze at me speechlessly. Finally, it occurred to me that he didn't understand English.

"Do you speak English?" I asked hopelessly.

"Of course!" he said vehemently. "What's wrong with you?"

"Something's wrong with something," I exclaimed. "I haven't any idea where I am or how I got here."

"Synthetic wine?" he asked sympathetically.

"Oh, hell! Think I'm a fool? Say, do you have a good man in mathematical physics on the faculty? Take me to him."

"Psychology, I should think," he said, studying me. "Or psychiatry. But I'm a law student and know nothing of either."

"Then make it mathematical physics, and I'll be grateful to you."

So I was conducted to the mathematical physicist. The student led me into the very building that corresponded to Morton Hall, and into an office the position of which corresponded to that of Woleshensky's office. However, the office was older and dustier; it had a Victorian look about it and was not as modern as Woleshensky's room. Professor Vibens was a rather small, bald-headed man with a keen-looking face. As I thanked the law student and started on my story, he looked rather bored, as though wondering why I had picked on him with my tale of wonder. Before I had gotten very far, he straightened up a little; and farther along he pricked up another notch; and before many minutes he was tense in his chair as he listened to me. When I finished, his comment was terse, like that of a man accustomed to thinking accurately and to the point.

"Obviously you come into this world from another set of coordinates. As we are on the z dimension, you must have come to us from the t dimension—".

He disregarded my attempts to protest at this point.

"Your man Woleshensky has evidently developed the conception of relativity further than we have, although Monpeters's theory comes close enough to it. Since I have no idea how to get you back, you must be my guest. I shall enjoy hearing all about your world."

"That is very kind of you," I said gratefully. "I'm accepting because I can't see what else to do. At least until the time when I can find a place in your world or get back to my own. Fortunately," I added as an afterthought, "no one will miss me there, unless it be a few classes of students who will welcome the little vacation that must elapse before my successor is found."

Breathlessly eager to discover what sort of a world I had gotten into, I walked with him to his home. And I may state at the outset that if I had found everything upside down and outlandishly bizarre, I should have been far less amazed and astonished than I was. For, from the walk that first evening from Professor Vibens's office along several blocks of residence street to his solid and respectable home, through all of my goings about the town and country during the years that I remained in the t-dimensional world, I found people and things thoroughly ordinary and familiar. They looked and acted as we do, and their homes and goods looked like ours. I cannot possibly imagine a world and a people that could be more similar to ours without actually being the same. It was months before I got over the idea that I had merely wandered into an unfamiliar part of my own city. Only the actual experience of wide travel and much sightseeing, and the knowledge that there was no such extensive English-speaking country in the world that I knew, convinced me that I must be on some other world, doubtless in the t dimension.

"A gentleman who has found his way here from another universe," the professor introduced me to a strapping young fellow mowing the lawn.

The professor's son was named John. Could anything be more commonplace?

"I'll have to take you around and show you things tomorrow," John said cordially, accepting the account of my arrival without surprise.

A redheaded servant girl, roast pork and rhubarb sauce for dinner, and checkers afterward, a hot bath at bedtime, the ringing of a telephone somewhere else in the house—is it any wonder that it was months before I would believe I had actually come into a different universe? What slight differences there were in the people and the world merely served to emphasize the similarity. For instance, I think they were just a little more hospitable and "old-fashioned" than we are. Making due allowances for the fact that I was a rather remarkable phenomenon, I think I was welcomed more heartily into this home and in others later; people spared me more of their time and interest from their daily business than would have happened under similar circumstances in a correspondingly busy city in America.

Again, John found a lot of time to take me about the city and show me banks and stores and offices. He drove a little squat car with tall wheels, run by a spluttering gasoline motor. (The car was not as perfect as our modern cars, and horses were quite numerous in the streets. Yet John was a busy businessman, the district superintendent of a life-insurance agency.) Think of it! Life insurance in Einstein's t dimension.

"You're young to hold such an important position," I suggested.

"Got started early," John replied. "Dad is disappointed because I didn't see fit to waste time in college. Disgrace to the family, I am."

What in particular shall I say about the city? It might have been any one of a couple of hundred American cities. Only it wasn't. The electric street cars, except for their bright green colour, were perfect; they might have been brought over from Oshkosh or Tulsa. The ten-cent stores with gold letters on their signs; drugstores with soft drinks; a mad, scrambling stock exchange; the blaring sign of an advertising dentist; brilliant entrances to motion-picture theatres were all there. The beauty shops did wonders to the women's heads, excelling our own by a good deal, if I am any judge; and at that time I had nothing more important on my mind than to speculate on that question. Newsboys bawled the *Evening Sun* and the *Morning Gale*, in whose curious flat type I could read accounts of legislative doings, murders, and divorces, quite as fluently as I could in my own *Tribune* at home. Strangeness and unfamiliarity had bothered me a good deal on a trip to Quebec a couple of years before; but they were not noticeable here in the t dimension.

For three or four weeks the novelty of going around, looking at things, meeting people, visiting concerts, theatres, and department stores was sufficient to absorb my interest. Professor Vibens's hospitality was so sincerely extended that I did not hesitate to accept, though I assured him that I would repay it as soon as I got established in this world. In a few days I was thoroughly convinced that there was no way back home. Here I must stay until I learned as much as Woleshensky knew about crossing dimensions. Professor Vibens eventually secured for me a position at the university.

It was shortly after I had accepted the position as instructor in experimental physics and had begun to get broken into my work that I noticed a strange commotion among the people of the city. I have always been a studious recluse, observing people as phenomena rather than participating in their activities. So for some time I noted only in a subconscious way the excited gathering in groups, the gesticulations and blazing eyes, the wild sale of extra editions of papers, the general air of disturbance. I even failed to take an active interest in these things when I made a railroad journey of three hundred miles and spent a week in another city; so thoroughly at home did I feel in this world that when the advisability arose of my studying laboratory methods at another university, I made the trip alone. So absorbed was I in my laboratory problems that I only noted the with half an eye the commotion and excitement everywhere, and merely recollected it later. One night it suddenly popped into my head that the country was aroused over something.

That night I was with the Vibens family in their living room. John tuned in the radio. I wasn't listening to the thing very much; I had troubles of my own. $F = g\, m_1 m_2 / r^2$ was familiar enough to me. It meant the same and held as rigidly here as in my old world. But what was the name of the bird who had formulated that law? Back home, it was Newton. Tomorrow in class I would have to be thoroughly familiar with his name. Pasvieux, that's what it was.

What messy surnames. It struck me that it was lucky that they expressed the laws of physics in the same form and even in the algebraical letters, or I might have had a time getting them confused— when all of a sudden the radio blatantly bawled: "THE GOSTAK DISTIMS THE DOSHES!"

John jumped to his feet.

"Damn right!" he shouted, slamming the table with his fist.

Both his father and mother annihilated him with withering glances, and he slunk from the room. I gazed stupefied. My stupefaction continued while the professor shut off the radio and both of them excused themselves from my presence. Then suddenly I was alert.

I grabbed a bunch of newspapers, having seen none for several days. Great sprawling headlines covered the front pages:

THE GOSTAK DISTIMS THE DOSHES.

For a moment I stopped, trying to recollect where I had heard those words before. They recalled something to me. Ah, yes! That very afternoon there had been a commotion beneath my window on the university campus. I had been busy checking over an experiment so that I might be sure of its success at tomorrow's class, and looked out rather absently to see what was going on. A group of young men from a dismissed class was passing and had stopped for a moment.

"I say, the gostak distims the doshes!" said a fine-looking young fellow. His face was pale and strained.

The young man facing him sneered derisively, "Aw, your grandmother! Don't be a feeble—".

He never finished. The first fellow's fist caught him in the cheek. Several books dropped to the ground. In a moment the two had clinched and were rolling on the ground, fists flying up and down, smears of blood appearing here and there. The others surrounded them and for a moment appeared to enjoy the spectacle, but suddenly recollected that it looked rather disgraceful on a university campus and, after a lively tussle separated the combatants. Twenty of them, pulling in two directions, tugged them apart.

The first boy strained in the grasp of his captors; his white face was flecked with blood, and he panted for breath.

"Insult!" he shouted, giving another mighty heave to get free. He looked contemptuously around. "The whole bunch of you ought to learn to stand up for your honour. The gostak distims the doshes!"

That was the astonishing incident that these words called to my mind. I turned back to my newspapers.

"Slogan Sweeps the Country," proclaimed the subheads. "Ringing Expression of National Spirit! Enthusiasm Spreads Like Wildfire! The new patriotic slogan is gaining ground rapidly," the leading article went on. "The fact that it has covered the country almost instantaneously seems to indicate that it fills a deep and long-felt want in the hearts of the people. It was first uttered during a speech in Walkingdon by that majestic figure in modern statesmanship, Senator Harob. The beautiful sentiment, the wonderful emotion of this sublime thought, are epoch-making. It is a great conception, doing credit to a great man, and worthy of being the guiding light of a great—".

That was the gist of everything I could find in the papers. I fell asleep still puzzled about the time. I was puzzled because—as I now see and didn't then—I was trained in the analytical methods of physical science and knew little or nothing about the ways and emotions of the masses of the people.

In the morning the senseless expression popped into my head as soon as I awoke. I determined to waylay the first member of the Vibens family who showed up, and demand the meaning of the thing. It happened to be John.

"John, what's a gostak?"

John's face lighted up with pleasure. He threw out his chest and a look of pride replaced the pleasure. His eyes blazed, and with a consuming enthusiasm he shook hands with me, as deacons shake hands with a new convert—a sort of glad welcome.

"The gostak!" he exclaimed. "Hurray for the gostak!"

"But what is a gostak?"

"Not a gostak! The gostak. The gostak is—the distimmer of the doshes—see! He distims 'em, see?"

"Yes, yes. But what is distimming? How do you distim?"

"No, no! Only the gostak can distim. The gostak distims the doshes. See?"

"Ah, I see!" I exclaimed. Indeed, I pride myself on my quick wit. "What are doshes? Why, they are the stuff distimmed by the gostak. Very simple!"

"Good for you!" John slapped my back in huge enthusiasm. "I think it wonderful for you to understand us so well after being here only a short time. You are very patriotic."

I gritted my teeth tightly to keep myself from speaking.

"Professor Vibens, what's a gostak?" I asked in the solitude of his office an hour later.

He looked pained.

"Hush!" he whispered. "A scientific man may think what he pleases, but if he says too much, people may misjudge him. As a matter of fact, a good many scientific men are taking this so-called patriotism seriously. But a mathematician cannot use words loosely; it has become second nature with him to inquire closely into the meaning of every term he uses."

"Well, doesn't that jargon mean anything at all?" I was beginning to be puzzled in earnest.

"To me, it does not. But it seems to mean a great deal to the public in general. It's making people do things, is it not?"

I stood a while in stupefied silence. That an entire great nation should become fired up over a meaningless piece of nonsense! Yet the astonishing thing was that I had to admit that there was plenty of precedent for it in the history of my z-dimensional world. A nation exterminating itself in civil wars to decide which of two profligate royal families should be privileged to waste the people's substance from the throne; a hundred thousand crusaders marching to death for an idea that to me means nothing; a meaningless, untrue advertising slogan that sells millions of dollars' worth of cigarettes to a nation, to the latter's own detriment—haven't we seen it over and over again?

"There's a public lecture on this stuff tonight at the First Church of The Salvation," Professor Vibens suggested.

"I'll be there," I said. "I want to look into the thing."

That afternoon there was another flurry of "extras" over the street; people gathered in knots and gesticulated with open newspapers.

"War! Let 'em have it!" I heard men shout.

"Is our national honour a rag to be muddied and trampled on?" the editorial asked.

As far as I could gather from reading the papers, there was a group of nations across an ocean that was not taking the gostak seriously. A ship whose pennant bore the slogan had been refused entrance to an Engtalian Harbor because it flew no national ensign. The Executive had dispatched a diplomatic note. An evangelist who had attempted to preach the gospel of the distimmed doshes at a public gathering in Itland had been ridden on a rail and otherwise abused. The Executive was dispatching a diplomatic note.

Public indignation waxed high. Derogatory remarks about "wops" were flung about. Shouts of "Holy war!" were heard. I could feel the tension in the atmosphere as I took my seat in the crowded church in the evening. I had been assured that the message of the gostak and the doshes would be thoroughly expounded so that even the most simple-minded and uneducated people could understand it fully. Although I had my hands full at the university, I was so puzzled and amazed at the course events were taking that I determined to give the evening to finding out what the "slogan" meant.

There was a good deal of singing before the lecture began. Mimeographed copies of the words were passed about, but I neglected to preserve them and do not remember them. I know there was one solemn hymn that reverberated harmoniously through the great church, a chanting repetition of "The Gostak Distims the Doshes." There was another stirring martial air that began, "Oh, the Gostak! Oh, the Gostak!" and ended with a swift cadence on "The Gostak Distims the Doshes!" The speaker had a rich, eloquent voice and a commanding figure. He stepped out and bowed solemnly.

"The gostak distims the doshes," he pronounced impressively. "Is it not comforting to know that there is a gostak; do we not glow with pride because the doshes are distimmed? In the entire universe there is no more profoundly significant fact: the gostak distims the doshes. Could anything be more complete, yet more tersely emphatic! The gostak distims the doshes!" Applause. "This thrilling truth affects our innermost lives. What would we do if the gostak did not distim the doshes? Without the gostak, without doshes, what would we do? What would we think? How would we feel?" Applause again.

At first, I thought this was some kind of an introduction. I was inexperienced in listening to popular speeches, lectures, and sermons. I had spent most of my life in the study of physics and its accessory sciences. I could not help trying to figure out the meaning of whatever I heard. When I found none, I began to get impatient. I waited some more, thinking that soon he would begin on the real explanation. After thirty minutes of the same stuff as I have just quoted, I gave up trying to listen. I just sat and hoped he would soon be through. The people applauded and grew more excited. After an hour I stirred restlessly; I slouched down in my seat and sat up by turns. After two hours I grew desperate; I got up and walked out. Most of the people were too excited to notice me. Only a few of them cast hostile glances at my retreat.

The next day the mad nightmare began for me. First there was a snowstorm of extras over the city, announcing the sinking of a merchantman by an Engtalian cruiser. A dispute had arisen between the officers of the merchantman and the port officials, because the latter had jeered disrespectfully at the gostak. The merchantman picked up and started out without having fulfilled all the custom requirements. A cruiser followed it and ordered it to return. The captain of the merchantman told them that the gostak distims the doshes, whereupon the cruiser fired twice and sank the merchantman. In the afternoon came the extras announcing the Executive's declaration of war.

Recruiting offices opened; the university was depleted of its young men; uniformed troops marched through the city, and railway trains full of them went in and out. Campaigns for raising war loans; home guards, women's auxiliaries, ladies' aid societies making bandages, young women enlisting as ambulance drivers—it was indeed war; all of it to the constantly repeated slogan: "The gostak distims the doshes."

I could hardly believe that it was really true. There seemed to be no adequate cause for a war. The huge and powerful nation had dreamed a silly slogan and flung it in the world's face. A group of nations across the water had united into an alliance, claiming they had to defend themselves against being forced upon a principle they did not desire. The whole thing at the bottom had no meaning. It did not seem possible that there would actually be a war; it seemed more like going through a lot of elaborate play-acting.

Only when the news came of a vast naval battle of doubtful issue, in which ships had been sunk and thousands of lives lost, did it come to me that they meant business. Black bands of mourning appeared on sleeves and in windows. Reports of a division wiped out by an airplane attack; of forty thousand dead in a five-day battle; of more men and more money needed, began to make things look real. Haggard men with bandaged heads and arms in slings appeared on the streets, a church and an auditorium were converted into hospitals, and trainloads of wounded were brought in. To convince myself that this thing was so, I visited these wards and saw with my own eyes the rows of cots, the surgeons working on ghastly wounds, the men with a leg missing or a hideously disfigured face.

Food became restricted; there was no white bread, and sugar was rationed. Clothing was of poor quality; coal and oil were obtainable only on government permit. Businesses were shut. John was gone; his parents received news that he was missing in action.

Real it was; there could be no more doubt of it. The thing that made it seem most real was the picture of a mangled, hopeless wreck of humanity sent back from the guns, a living protest against the horror of war. Suddenly someone would say: "The gostak distims the doshes!" and the poor wounded fragment would straighten up and put out his chest with pride, and an unquenchable fire would blaze in his eyes. He did not regret having given his all for that. How could I understand it?

And real it was when the draft was announced. More men were needed; volunteers were insufficient. Along with the rest, I complied with the order to register, doing so in a mechanical fashion, thinking little of it. Suddenly the coldest realization of the reality of it was flung at me when I was informed that my name had been drawn and that I would have to go!

All this time, I had looked upon this mess as something outside of me, belonging to a different world, of which I was not a part. With all this death and mangled humanity in the background, I wasn't even interested in this world. I didn't belong here. I didn't want to undergo all the horrors of military life, the risk of a horrible death for no reason at all! For a silly jumble of meaningless sounds.

I spent a sleepless night in maddened shock from the thing. In the morning a wild and haggard caricature of myself looked back at me from the mirror. But I had revolted. I intended to refuse service. If the words "conscientious objector" ever meant anything, I certainly was one. Even if they shot me for treason at once, that would be a fate less hard to bear than going out and giving my strength and my life for—for nothing at all.

My apprehensions were quite correct. With my usual success at self-control over a seething interior, I coolly walked to the draft office and informed them that I did not believe in their cause and could not see my way to fight for it. Evidently, they had suspected something of that sort already, for they had the irons on my wrists before I had hardly done with my speech.

"Period of emergency," said a beefy tyrant at the desk. "No time for stringing out a civil trial. Court-martial!"

He said it to me vindictively, and the guards jostled me roughly down the corridor; even they resented my attitude. The court-martial was already waiting for me. From the time I walked out of the lecture at the church I had been under secret surveillance, and they knew my attitude thoroughly. This is the first thing the president of the court informed me.

My trial was short. I was informed that I had no valid reason for objecting. Objectors because of religion, because of nationality and similar reasons, were readily understood; a jail sentence to the end of the war was their usual fate. But I had admitted no intrinsic objection to fighting; I merely jeered at their holy cause. That was treason unpardonable.

"Sentenced to be shot at sunrise!" the president of the court announced.

The world spun around me. But only for a second. My self-control came to my aid. With the curious detachment that comes to us in such emergencies I noted that the court-martial was being held in Professor Vibens's office—that dingy little Victorian room where I had first told my story of travelling by relativity and had first realized that I had come to the t-dimensional world. Apparently, it was also to be the last room I was to see in this same world. I had no false hopes that the execution would help me back to my own world, as such things sometimes do in stories. When life is gone, it is gone, whether in one dimension or another. I would be just as dead in the z dimension as in the t dimension.

"Now, Einstein, or never!" I thought. "Come to my aid, O Riemann! O Lobachevski! If anything will save me it will have to be a tensor or a geodesic."

I said it to myself rather ironically. Relativity had brought me here. Could it get me out of this?

Well! Why not?

If the form of a natural law, yea, if a natural object varies with the observer who expresses it, might not the truth and the meaning of the gostak slogan also be a matter of relativity? It was like making the moon ride the treetops again. If I could be a better relativist and put myself in these people's places, perhaps I could understand the gostak. Perhaps I would even be willing to fight for him or it.

The idea struck me suddenly. I must have straightened up and some bright change must have passed over my features, for the guards who led me looked at me curiously and took a firmer grip on me. We had just descended the steps of the building and had started down the walk.

Making the moon ride the treetops! That was what I needed now. And that sounded as silly to me as the gostak. And the gostak did not seem so silly. I drew a deep breath and felt very much encouraged. The viewpoint of relativity was somehow coming back to me. Necessity manages much. I could understand how one might fight for the idea of a gostak distimming the doshes. I felt almost like telling these men. Relativity is a wonderful thing. They led me up the slope, between the rows of poplars.

Then it all suddenly popped into my head: how I had gotten here by changing my coordinates, insisting to myself that I was going upward. Just like making the moon stop and making the trees ride when you are out riding at night. Now I was going upward. In my world, in the z dimension, this same poplar was down the slope.

"It's downward!" I insisted to myself. I shut my eyes and imagined the building behind and above me. With my eyes shut, it did seem downward. I walked for a long time before opening them. Then I opened them and looked around.

I was at the end of the avenue of poplars. I was surprised. The avenue seemed short. Somehow it had become shortened; I had not expected to reach the end so soon. And where were the guards in olive uniform? There were none.

I turned around and looked back. The slope extended backward above me. Indeed, I had walked downward. There were no guards, and the fresh, new building was on the hill behind me.

Woleshensky stood on the steps.

"Now, what do you think of a t dimension?" he called out to me.

Woleshensky!

And a new building, modern! Vibens's office was in an old Victorian building. What was there in common between Vibens and Woleshensky? I drew a deep breath. The comforting realization spread gratefully over me that I was back in my native dimension. The gostak and the war were somewhere else. Here were peace and Woleshensky.

I hastened to pour out the story to him.

"What does it all mean?" I asked when I was through. "Somehow—vaguely—it seems that it ought to mean something."

"Perhaps," he said in his kind, sage way, "we really exist in four dimensions. A part of us and our world that we cannot see and are not conscious of projects on into another dimension, just like the front edges of the books in the bookcase, turned away from us. You know that the section of a conic cut by the *y* plane looks different than the section of the same conic by the *z* plane? Perhaps what you saw was our world and our own selves intersected by a different set of coordinates. Relativity, as I told you in the beginning."

> **Miles John Breuer** (1889–1945) was an American medical doctor who published a number of science fiction tales in pulp magazines such as "Amazing Stories" and "Astounding Stories". Several of Breuer's narratives had their roots in ideas from mathematics and physics. He collaborated on a novel "The Girl from Mars" with the much better known science fiction writer Jack Williamson.

Commentary

After reading the commentary to Chap. 1, in which we discussed the three dimensions of space and the possibility of higher dimensions, anyone with a basic background in science might well ask: "What about time? Time is a dimension too, right?" Well, yes, time is a dimension. And time and space dimensions are, in a sense, interchangeable. That much Breuer got right in his story. But time and space display important differences. "Changing from the z to the t coordinate", as Professor Woleshensky suggests doing in Breuer's story, is not something we can do at will in our universe. And the differences between time and space all depend upon a minus sign. Why? Let's see.

First, let's remind ourselves, as if we needed reminding, that the dimensions of time and space *feel* different. We can go backwards and forwards in space while we must travel in a single direction in time. More than that, we find ourselves *situated* in space whereas time possesses an aspect of *flow*. If you go for a walk and come across a boulder then you don't necessarily expect to see more boulders in the space around you. In the time dimension, though, you expect to see the boulder at the next instant, and the next, and the next. Time *feels* different.

We must take care, though, with mere feelings. As the special theory of relativity tells us, our everyday experience of time and space can lead us astray. Einstein argued the universe possesses an absolute speed limit, a speed upon which all observers agree. That limit defines the speed at which fundamental interactions can propagate. One of those fundamental interactions is electromagnetism or, to use the common term, light. The existence of a universal speed limit, then, defines the speed at which light can travel through empty space. Einstein thus argued that all observers, no matter how fast they move, agree on the speed of light. That principle, in turn, forces us adapt our notions of time and space. Einstein's former teacher, Hermann Minkowski, told us we should not think of 'space' and 'time' as separate entities. Rather, we should think of a unified, four-dimensional 'spacetime'.

Minkowski spacetime, which springs from Einstein's ideas of special relativity, is flat; static; infinite. (Einstein later realised spacetime can be curved; dynamic; infinite *or* finite. In a stroke of genius, he identified curved spacetime with gravity. This identification lies at the heart of his theory of general relativity. We'll return to this later, but for the moment let's stick with thinking about flat spacetime.) The properties of Minkowski spacetime explain why we cannot swap between space and time in the manner suggested by Professor Woleshensky in Breuer's story.

According to Minkowski, an 'event' in spacetime is a point in the universe that requires four coordinates to locate it. Three coordinates fix the event in space, one coordinate fixes the event in time. Suppose event A is: leave your home at 7 a.m. And suppose event B is: arrive at office desk at 8 a.m. We can ask two questions about events A and B, the answers to which lay bare the difference between space and time.

First question: what is the distance in space between events A and B? Well, that depends upon the route taken. In a flat spacetime, the shortest distance between two points is a straight line. And we can write down an equation that gives us that distance. Since Breuer included an equation in his story, I feel I have permission to present an equation here. (In any case, I am sure you will already have seen the equation.)

Let's begin with some terminology.

Let's agree to use the Greek capital letter Delta, symbol Δ, to denote the phrase "the amount of change in". Let's further agree to use the more familiar symbol s to represent the separation between two points. The combined symbol Δs thus means "the amount of change in s".

If we focus on the separation between two points in the x direction then we could also use the symbol Δx. In one dimension we have the trivial equation $\Delta s = \Delta x$, since in one dimension s and x represent the same quantity. If we wanted to, we could square both sides of the equation and write it as $(\Delta s)^2 = (\Delta x)^2$. That might seem like a pointless thing to do, but a good reason for doing so will soon become clear.

Now let's look at two dimensions. Represent the separation between points A and B in the x direction by the symbol Δx. And represent the separation between the same points in the y direction by the symbol Δy. Then the straight line separation Δs between A and B is given by the Pythagorean theorem: $(\Delta s)^2 = (\Delta x)^2 + (\Delta y)^2$.

The same form of the equation holds in three dimensions. We just have to include the separation between A and B in the z direction. The Pythagorean theorem in three dimensions gives us: $(\Delta s)^2 = (\Delta x)^2 + (\Delta y)^2 + (\Delta z)^2$.

Look at the equations for one, two, and three dimensions. The form of the equation holds in each case: the square of the separation between two points equals the sum of the squares of the separations in each of the different dimensions. Simple!

Except … in the real world any trip between two points almost inevitably covers more ground than the straight-line distance. If you walk, you take a route determined by the provision of sidewalks. If you drive, the roads determine your path. If you take public transport, you must get to and from pick-up points. None of those options necessarily lie on the shortest, straight-line distance. Different observers can all start and end at the same points in space but if they follow different paths between those points they will measure different distances. And those distances will be *longer* than the straight-line distance.

Second question: what time elapses between event A and event B? This seems straightforward: the elapsed time is one hour. A occurs at 7 a.m. and B occurs at 8 a.m. But this assumes everyone agrees on the time coordinate in spacetime. That's not the case. Relativity tells us observers in relative motion make different measurements of time.

If you carry a clock with you as you travel between A and B then you can measure the "proper time" between these two events. Your clock's reading gives you the proper time between events in the same way your car's odometer gives you the spatial distance between events. But both measurements—of time and space—depend upon the path taken.

To travel along a straight path in spacetime means following a straight line in space while moving at a constant velocity. Here comes the key difference between time and space. The straightest path between two events in spacetime gives the *longest elapsed time* between those events while it gives the *shortest distance* between those events. This deserves repeating. The separation between two events on a straight path in Minkowski spacetime records the *shortest distance* and the *longest time*. If you take any other path through spacetime between those two events then you travel a *greater* distance but measure a *shorter* time.

We can write the Pythagoras equation in Minkowski spacetime, and it makes clear the difference between time and space.

The spacetime interval Δs between two events, separated by Δx, Δy, Δz in space and by Δt in time, obeys the following equation: $(\Delta s)^2 = -(\Delta t)^2 + (\Delta x)^2 + (\Delta y)^2 + (\Delta z)^2$. Notice the critical sign appearing before the Δt term in that equation. The space coordinates have a *plus* sign beside them; the time coordinate has a *minus* sign beside it. (Some people prefer to write this with a minus sign for the space coordinates and a plus sign for the time coordinate. Either way produces the same results.) That sign difference accounts for the huge difference in how time and space work.

Actually, we must amend the equation above. Just slightly. The units don't work. We measure time in seconds, distances in light-seconds. (At least, we *would* measure distances in light-seconds if we were relativistic creatures. Since we inhabit a low-speed world, where relativistic effects go unnoticed, we measure distances in metres or miles or something.) Since we use different units when measuring time and space, we need a constant in the equation to make the units agree. That constant is the speed of light, c. So the equation for the spacetime interval between two events becomes: $(\Delta s)^2 = -(c\Delta t)^2 + (\Delta x)^2 + (\Delta y)^2 + (\Delta z)^2$. If we work in units in which c is 1 light-second per second, then we can write $c = 1$ and the first equation stands.

We know of no deep reason (at least, none that I know) for *why* the sign difference exists. The sign difference just reflects the workings of our universe and the baking-in of causality. That sign difference gives rise to all the seemingly peculiar effects of special relativity: time dilation, length contraction, the twin paradox, and so on. But the sign difference means a switch between the z and t coordinate, as in Breuer's tale, is not something one can do at will. That's not how the universe works.

The story, though, doesn't end there.

The above argument applies to flat Minkowski spacetime. The presence of mass curves spacetime. When we talk about stars, or galaxies, or the universe on a large scale, we need to take account of spacetime curvature. To calculate the interval between two events in curved spacetime, we need to use Einstein's general theory of relativity. Without going into details, it turns out that at certain places, such as in the interior of black holes, there *is* a sense in which time and space switch roles. Sort of. Imagine an astronaut who has crossed the event horizon of a black hole. For that unfortunate individual, the space coordinates giving the distance to the singularity inevitably *decrease*—just as the time coordinate inevitably *increases* for those of us living in flat spacetime. For that stranded astronaut, the black hole singularity lies not "over there" but rather "in the future".

Breuer did not have the spacetime curvature of black holes on his mind when developing his story. I doubt he was even thinking about flat Minkowski spacetime when writing the story. Rather, for Breuer, the swapping between z and t coordinates offered a way to develop his political satire. But his satirical point, I think, was well made. Have we not seen similar events in recent years, with political parties in the UK and US spreading slogans that make as much sense—and have the same divisive potential—as "the gostak distims the doshes"?

Notes and Further Reading

As the special theory of relativity tells us—In 1905, Albert Einstein (1879–1955) wrote a paper introducing concepts of special relativity and another paper describing a famous consequence of the theory, the equivalence of mass and energy. These were two of his four *annus mirabilis* papers. For a thorough yet accessible introduction to special relativity, see Susskind and Friedman (2018).

we need to use Einstein's general theory of relativity—Einstein developed his general theory of relativity between the years 1907 to 1915. General relativity is more mathematically involved than special relativity. Susskind and Cabannes (2024), a sequel to the volume on the special theory, provides a good introduction to general relativity.

References

Susskind, L., Cabannes, A.: General Relativity: The Theoretical Minimum. Penguin, London (2024)

Susskind, L., Friedman, A.: Special Relativity and Classical Field Theory: The Theoretical Minimum. Penguin, London (2018)

3

Messaging Extraterrestrial Intelligence

Love and a Triangle (Stanley Waterloo)

A man came out of a mine, looked about him, inhaled the odour from the stunted spruce trees, looked up at the clear skies, then called to a boy idling in a shed at a little distance from the mine buildings, telling him to bring out the horse and buckboard. The name of the man who had issued from the mine was Julius Corbett, and he was a civil engineer. Furthermore, he was a capitalist.

 He was an intelligent looking man of about thirty-five, and a resolute looking one, this Julius Corbett, and as he stood waiting for the buckboard, was rather worth seeing, vigorous of frame, clear of eye and bronzed by a summer's work in a wild country. The shaft from which he had just emerged was that of a silver mine not five miles distant from Black Bay, one of the inlets of the northern shore of Lake Superior, and was a most valuable property, of which he was chief owner. He had inherited from an uncle in Canada a few hundred acres of land in this region, but had scarcely considered it worthy the payment of its slight taxes until some of the many attempts at mining in the region had proved successful, and it was shown that the famous Silver Islet, worked out years ago in Lake Superior, was not the only repository thereabouts of the precious metal. Then he had abandoned for a time the practice of his profession—he had an office in Chicago—and had visited what he referred to lightly as his "British possessions". He had found rich indications, had called in mining experts, who confirmed all he had imagined, and had returned to Chicago and organized a company. There was a monotonous success to the undertaking, much at variance with the story of ordinary mining enterprises. Corbett had become a very rich man within two years; he was worth more than a million, and was becoming richer daily. He was, seemingly, a person much to be envied, and would not himself, on the day here referred to, have denied such imputation, for he was in love with an exceedingly sweet and clever girl, and knew that he had won this same charming creature's heart. They were plighted to each other, but the date of their marriage was not yet fixed. He had closed up his business at the mine for the season, and was now about to hasten to Chicago, where the day of so much importance to him would be fixed upon and the sum of his good fortune soon made complete. This was in September, 1898.

It was not a commonplace girl whom Corbett was to marry. On the contrary, she was exceptionally gifted, and a young woman whose cleverness had been supplemented by an elaborate education. There was, however, running through her character a vein of what might be called emotionalism. The habit of concentration, acquired through study, seemed rather to intensify this quality than otherwise. Perhaps it made even greater her love for Corbett, but it was destined to perplex him.

In September the air is crisp along the route from Black Bay to Duluth, and from that through fair Wisconsin to Chicago, and Corbett's spirits were high throughout the journey. Was he not to meet Nell Morrison, in his estimation the sweetest girl on earth? Was he not soon to possess her entirely and for a permanency? He made mental pictures of the meeting, and drifted into a lover's mood of planning. Out of his wealth what a home he would provide for her, and how he would gratify her gentle whims! Even her astronomical fancy, Vassar-born, should become his own, and there should be an observatory to the house. He had a weakness for astronomy himself, and was glad his wife-to-be had the same taste intensified. They would study the heavens together from a heaven of their own. What was wealth good for anyhow, save to make happy those we love?

The train sped on, and Chicago was reached, and very soon thereafter was reached the home of the Morrisons. Corbett could not complain of his reception. The one creature was there, sweet as a woman may be, eager to meet him, and with tenderness and steadfastness shown in every line of her pretty face. They spent a charming day and evening together, and he was content. Once or twice, just for a moment, the young woman seemed abstracted, but it was only for a moment, and the lover thought little of the circumstance. He was happy when he bade her good-night. "Tomorrow, dear," said he, "we will talk of something of greatest importance to me, of importance to us both." She blushed and made no answer for a second. Then she said that she loved him dearly, and that what affected one must affect the other, and that she would look for him very early in the afternoon. He went to his hotel buoyant. The world was good to him.

When Corbett called at the Morrison mansion the next day he entered without ringing, as was his habit, and went straight to the library, expecting to find Nell there. He was disappointed, but there were traces of her recent presence. There was an astronomical map open upon the table, and books and reviews lay all about, each, open, with a marker indicating a special page. A little glove lay upon the floor, and Corbett picked it up and kissed it.

He summoned a servant and sent upstairs to announce his presence; then turned instinctively to note what branch of her favourite study was now attracting his sweetheart's attention. He picked up one of the open reviews, an old one by the way, and read a marked passage there. It was as follows:

"It will always be more difficult for us to communicate with the people of Mars than to receive signals from them, because of our position and phases. It is the nocturnal terrestrial hemisphere that is turned toward the planet Mars in the periods when we approach most nearly to it, and it shows us in full its lighted hemisphere. But communication is possible."

He looked at a map. It was a great chart of the surface of Mars, made by the famous Italian Schiaparelli, and he looked at more of the reviews and found ever the same subject considered in the marked articles. All related to Mars. He was puzzled but delighted. "The dear girl has a hobby," he thought. "Well, she shall enjoy it to the utmost."

Nelly entered the room. Her face lighted up with pleasure when she met her fiancé, but assumed a more thoughtful look as she saw what he was reading. She welcomed him, though, as kindly as any lover could demand, and he, of course, was joyously content. "Still an astronomer, I see," he said, "and apparently with a specialty. I see nothing but Mars, all Mars! Have you become infatuated with a single planet, to the neglect of all the others? I like it, though. We will study Mars together."

Her face brightened. "I am so glad!" she said. "I have studied nothing else for months. It has been so almost from the day you left us. And it is not Mars alone I am studying; it is the great problem of communication with the people there. Oh, Julius, it is possible, and the idea is something wonderful! Just think what would follow! It would be the beginning of an understanding between reasoning creatures of the whole universe!"

He said that it was something wonderful, indeed, maybe only a dream, but a very fascinating one.

"Oh, it is no dream," she answered. "It is a glorious possibility. Why, just think of it, we know, positively know, that Mars is inhabited. Think of what has been discovered. It was perceived years ago that Mars was intersected by canals, evidently made by human—I suppose that's the word—human beings. They run from the extremes of ocean bays to the extremes of other ocean bays, and connect, too, the many lakes there. Nature does not make such lines. They are of equal width, those canals, throughout their whole length, and Schiaparelli has even watched them in construction. First there is a dark line, as if the earth had been disturbed, and then it becomes bright when the water is let in. Sometimes, too, double canals are made there close to each other, running side by side, as if one were used for travel and transportation in one direction and one in another. And there are many other things as wonderful. The world of Mars is like our own. There are continents and seas and islands there—it is not a dead, dry surface like the moon—and it has clouds and rains and snows and seasons, just as we have, and of the same intensity as ours. Oh, Julius, we must communicate with them!"

"But, my dear, that implies equal interest on their part. How do we know them to be intelligent enough?"

"Why, there are the canals. They must be reasoners in Mars. Besides, how do we know but that they far surpass us in all learning! Mars is much older in one way than the Earth, far more advanced in its planet life, and why should not its people, through countless ages of advantage, have become wiser than we? Whatever their form, they may be superior to us in every way. We are to them, too, something which must have been studied for thousands of years. The Earth, you know, is to the people on Mars a most brilliant object. It is the most glorious object in their sky, a star of the first magnitude. Oh, be sure their astronomers are watching us with all interest!"

And Corbett, dazed, replied that he was overwhelmed with so much learning in one so fair, that he was very proud of her, but that there was one subject on his mind, compared to which communication with Mars or any other planet was but a trifle. And he wanted to talk with her concerning what was closest to his heart. It was the one great question in the world to him. It was, when should be their wedding day?

The girl looked at him blushingly, then paled. "Let us not talk of that today," she said, at length. "I know it isn't right; I know that I seem unkind—but—oh, Julius! come tomorrow and we will talk about it." And she began crying.

He could not understand. Her demeanour was all incomprehensible to him, but he tried to soothe her, and told her she had been studying too hard and that her nerves were not right. She brightened a little, but was still distrait. He left, with something in his heart like a vengeful feeling toward the planets, and toward Mars in particular.

When Corbett returned next day the girl was in the library awaiting him. Her demeanour did not relieve him. He feared something indefinable. She was sad and perplexed of countenance, but more self-possessed than on the day before. She spoke softly: "Now we will talk of what you wished to yesterday."

He pleaded as a lover will, pleaded for an early day, and gave a hundred reasons why it should be so, and she listened to him, not apathetically, but almost sadly. When he concluded, she said, very quietly:

"Did you ever read that queer story by Edmond About called 'The Man with the Broken Ear'?"

He answered, wonderingly, in the affirmative.

"Well, dear" she said, "do you remember how absorbed, so that it was a very part of her being, the heroine of that story became in the problem of reviving the splendid mummy? She forgot everything in that, and could not think of marriage until the test was made and its sequel satisfactory. She was not faithless; she was simply helpless under an irresistible influence. I'm afraid, love"—and here the tears came into her eyes—"that I'm like that heroine. I care for you, but I can think only of the people in Mars. Help me. You are rich. You have a million dollars, and will soon have more. Reach those people!"

He was shocked and disheartened. He pleaded the probable utter impracticability of such an enterprise. He might as well have talked to a statue. It all ended with an outburst on her part.

"Talk with the Martians," said she, "and the next day I will become your wife!"

He left the house a most unhappy man. What could he do? He loved the girl devotedly, but what a task had she given him! Then, later, came other reflections. After all, the end to be attained was a noble one, and he could, in a measure, sympathize with her wild desire. The lover in 'The Man with the Broken Ear' had at least occasion for a little jealousy. His own case was not so bad. He could not well be jealous of an entire population of a distant planet. And to what better use could a portion of his wealth be put than in the advancement of science! The idea grew upon him. He would make the trial!

He was rewarded the next day when he told his fiancée what he had decided upon. She was wildly delighted. "I love you more than ever now!" she declared, "and I will work with you and plan with you and aid you all I can. And," she added, roguishly, "remember that it is not all for my sake. If you succeed you will be famous all over the world, and besides, there'll come some money back to you. There is the reward of one hundred thousand francs left in 1892 by Madame Guzman to any one who should communicate with the people of another planet."

He responded, of course, that he was impelled to effort only by the thought of hastening a wedding day, and then he went to his office and wrote various letters to various astronomers. His friend Marston, professor of astronomy in the University of Chicago, he visited in person. He was not a laggard, this Julius Corbett, in anything he undertook.

Then there was much work.

Marston, being an astronomer, believed in vast possibilities. Being a man of sense, he could advise. He related to Corbett all that had been suggested in the past for interstellar communication. He told of the suggested advice of making figures in great white roads upon some of Earth's vast plains, but dismissed the idea as too costly and not the best. "We have a new agent now," he said. "There is electricity. We must use that. And the figures must, of course, be geometrical. Geometry is the same throughout all the worlds that are or have been or ever will be."

And there was much debate and much correspondence and an exhibition of much learning, and one day Corbett left Chicago. His destination was Buenos Ayres, South America.

The Argentine Republic, since its financial troubles early in the decade, had been in a complaisant and conciliating mood toward all the world, and Corbett had little difficulty in his first step—that of securing a concession for stringing wires in any designs which might suit him upon the vast pampas of the interior. It was but stipulated that the wires should be raised at intervals, that herding might not be interfered with. He had already made a contract with one of the great electric companies. The illuminated figures were to be two hundred miles each in their greatest measurement, and were to be as follows:

It was found advisable, later, to dispense with the last two, and so, only the square, equilateral triangle, circle and right-angled triangle, it was decided should be made. The work was hurried forward with all the impetus of native energy, practically unlimited money and the power of love. This last is a mighty force.

And great works were erected, with vast generators, and thousands and thousands of miles of sheets of wires were strung close together, until each system, when illuminated, would make a broad band of flame surrounding the defined area. From the darkened surface of the Earth, at the time when the Earth approached Mars most nearly, would blaze out to the Martians the four great geometrical figures. The test was made at last. All that had been hoped for in the way of an effort was attained. All along the lines of those great figures, night in the Argentine Republic was turned into glorious day. From balloons the spectacle was something incomparably magnificent. All was described in a thousand letters. A host of correspondents were there, and accounts of the undertaking and its progress were sent all over the civilized world. Each night the illumination was renewed, and all the world waited. Months passed.

Corbett had returned to Chicago. He could do no more. He could only await the passage of time, and hope. He was not very buoyant now. His sweetheart was full of the tenderest regard, but was in a condition of feverish unrest. He was alarmed regarding her, so great appeared her anxiety and so tense the strain upon her nerves. He could not help her, and prepared to return again to a season at his mine.

The man was sitting in his room one night in a gloomy frame of mind. What a fool he had been! He had but yielded to a fancy of a dreaming girl, and put her even farther away from him while wasting half a fortune! He would be better on the rugged shore of Lake Superior, where the moods of men were healthy, and where were pure air and the fragrance of the pines. There was a strong pull at his bell.

A telegraph boy entered, and this was on the message he bore:
Come to the observatory at once. Important.
MARSTON.

To seek a cab, to be whirled away at a gallop to the university, to burst into Marston in his citadel, required but little time. The professor was walking up and down excitedly.

"It has come! All the world knows it!" he shouted as Corbett entered, and he grasped him by the hand and wrung it hardly.

"What has come?" gasped the visitor.

"What has come, man! All we had hoped for or dreamed of—and more! Why, look! Look for yourself!"

He dragged Corbett to the eye-piece of the great telescope and made him look. What the man saw made him stagger back, overcome with an emotion which for the moment did not allow him speech. What he saw upon the surface of the planet Mars was a duplication of the glittering figures on the pampas of the South American Republic. They were in lines of glorious light, between what appeared bands of a darker hue, provided, apparently, to make them more distinct, and even at such vast distance, their effect was beautiful. And there was something more, a figure he could not comprehend at first, one not in the line of the others, but above. "What is it—that added outline?" he cried.

"What is it! Look again. You'll determine quickly enough! Study it!" roared out Marston, and Corbett did as he was commanded. Its meaning flashed upon him.

There, just above the representation of the right-angled triangle, shone out, clearly and distinctly, this striking figure:

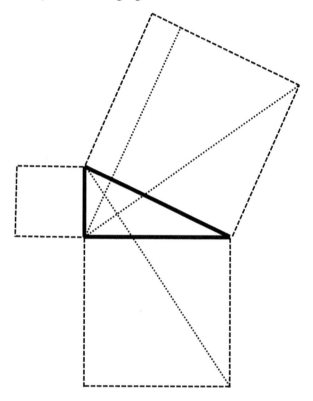

What could it mean? Ah, it required no profound mathematician, no veteran astronomer, to answer such a question! A schoolboy would be equal to the task. The man of Mars might have no physical resemblance to the man of Earth, the people of Mars might resemble our elephants or have wings, but the eternal laws of mathematics and of logic must be the same throughout all space. Two and two make four, and a straight line is the shortest distance between two points throughout the universe. And by adding this figure to the others represented, the Martians had said to the people of Earth as plainly as could have been done in written words of one of our own languages:

Yes, we understand. We know that you are trying to communicate with us, or with those upon some other world. We reply to you, and we show to you that we can reason by indicating that the square of the hypotenuse of a right-angled triangle is equivalent to the sum of the squares of the other two sides. Hope to hear from you further.

There was the right-angled triangle, its lines reproduced in unbroken brilliancy, and there were the added lines used in the familiar demonstration, broken at intervals to indicate their use. The famous pons asinorum had become the bridge between two worlds.

Corbett could scarcely speak as yet. Telegraph messengers came rushing in with dispatches from all quarters—from the universities of Michigan and California, and Yale and Harvard, and from Rochester and all over the United States. Cablegrams from England, France, Germany and Italy and other regions of the world but repeated the same wonderful observation, the same conclusion: "They have answered! We have talked with them!"

Corbett returned to his home in a semi-delirium. He had the wisdom, though it was midnight, to send to Nelly the brief message, "Good news", to prepare her in a degree for what the morning papers would reveal. He slept but fitfully. And it was at an early hour when he called upon his fiancée and found her awaiting him in the library.

She said nothing as he entered, but he had scarcely crossed the threshold when he found his arms full of something very tangible and warm, and pulsing with all love. It has been declared by thoughtful and learned people that there is no sensation in the world more delightful than may be produced by just this means, and Corbett's demeanour under the circumstances was such as to indicate the soundness of the assertion. He was a very happy man.

And she, as soon as she could speak at all, broke out, impulsively:

"Oh, dear, isn't it glorious! I knew you would succeed. And aren't you glad I imposed the hard condition? It was hard, I know, and I seemed unloving, but I believed, and I could not have given you up even if you had failed. I should have told you so very soon. I may confess that now. And—I will marry you any day you wish."

She blushed magnificently as she concluded, and the face of a pretty women, so suffused, is a pleasing thing to see.

Of course, within a week the name of Corbett became familiar in every corner of the civilized globe, the incentive which had spurred him on became somehow known, and the romance of it but added to his fame, and a few days later, when his wedding occurred, it was chronicled as never had a wedding been before. They made two columns of it even in the far-away Tokio Gazette, the Bombay Times and the Novgorod News. But the social feature was nothing; the scientific world was all aflame.

We had talked with Mars indeed, but of what avail was it if we could not resume the conversation? What next step should be taken in the grand march of knowledge, in the scientific conquest of the universe? Never in all history had there been such a commotion among the learned. Corbett and his gifted wife were early ranked among the eager, for he soon became as much of an enthusiast as she—in fact, since the baby, he is even more so—and derived much happiness from their mutual study and speculation. All theories were advanced from all countries, and suggestions, wise and otherwise, came from thousands of sources. And so in the year 1900 the thing remains. As inscrutable to us have been the curious symbols appearing upon Mars of late as have apparently been to them a sign language attempted on the pampas. It is now proposed to show to them the outline of a gigantic man, and if Providence has seen fit to make reasoning beings in all worlds something alike, this may prove another bit of progress in the intercourse, but all is in doubt.

Given, the problem of two worlds, millions of miles apart, the people of which are seeking to establish a regular communication with each other, each already acknowledging the efforts of the other, how shall the great feat be accomplished? Will the solution of the vast problem come from a greater utilization of electricity and a further knowledge of what is astral magnetism? There have been, of late, some wonderful revelations along that line. Or will the sign language be worked out upon the planets' surfaces? Who can tell? Certainly all effort has been stimulated, in one world at least. The rewards offered by various governments and individuals now aggregate over five million dollars, and all this money is as nothing to the fame awaiting some one. Who will gain the mighty prize? Who will solve the new problem of the ages?

> **Stanley Waterloo** (1846–1913) was an American editor and author who wrote a handful of novels with science fictional elements and a similar number of science fiction short stories. His most famous novel is "The Story of Ab: a Tale of the Time of the Cave Man", which proved to be such an influence on Jack London's tale "Before Adam" that Waterloo accused London of plagiarism.

Commentary

The Pythagorean theorem is one of those things we learn by rote in the schoolroom. As Stanley Waterloo tells it: "the square of the hypotenuse of a right-angled triangle is equal to the sum of the squares of the other two sides". In symbolic form, in two dimensions, the theorem takes a straightforward form: $(\Delta s)^2 = (\Delta x)^2 + (\Delta y)^2$. As we saw in the commentary to Chap. 2, the three-dimensional version of theorem takes the same simple form and we learned how to extend it to describe the interval between two events in Minkowski spacetime. Pythagoras, then, came up with a useful and important theorem. (Actually, some unknown Babylonian mathematician, living more than one thousand years before Pythagoras, proposed the equation first.) In "Love and a Triangle", Waterloo assumes *any* intelligent species will know about this theorem. He bases this assumption on the grounds that "the eternal laws of mathematics and of logic must be the same throughout all space". This common ground might allow us to make contact with extraterrestrial intelligence.

Even if we ignore the dated social attitudes, Waterloo's story can seem quaint to modern readers. The problem is not the science. At first glance, the story's basic thesis seems reasonable. Indeed, Waterloo was not the first to suggest we should use geometry to signal our presence to Mars. Carl Friedrich Gauss, one of the greatest of all mathematicians, suggested something similar. The underlying idea makes sense. Cultural references are local; mathematics, so far as we know, is universal. An alien species will have no knowledge of the Kardashians; but an intelligent species would understand the relationship between the sides of a right-angled triangle. (Or would they? Intelligent aquatic creatures, for example, might never encounter a triangle. Might their thinking instead revolve around more fluid concepts? But let's restrict ourselves to creatures with a similar evolutionary history to us.)

What makes the story quaint is not the notion of using mathematics to communicate with aliens, but rather the method suggested.

Waterloo split the problem of interplanetary communication into two parts: broadcast and reception. Let's start with the problem of reception.

For communication to take place, the putative Martian audience must detect our signal. We can assume they would use some form of telescope for this. When Waterloo wrote "Love and a Triangle", in 1899, the largest telescope on Earth was the Leviathan of Parsonstown, a reflector with a 72-inch mirror. Eighteen years later, the 100-inch Hooker Telescope at Mount Wilson would see first light. These are the sorts of instruments needed to observe the surface of Mars from Earth, or Earth from Mars. Well, let's assume Martian astronomers had developed large optical telescopes—just as we humans had done. Under that assumption, Martians could observe Earth.

The other part of the problem is: how do we send an optical signal to Mars? The Victorian scientist wishing to broadcast our existence to Martians had limited options. He (and the Victorian scientist was likely male, or else a male taking credit for a female's work) could dig a vast Pythagorean trench in the Sahara desert, fill the trench with oil, and set fire to it. Or he could plant a gigantic forest with noticeable geometric boundaries in Siberia. Or he could do as the character does in this story. Or he … well, that about exhausts the options. Available technology constrains possibilities (and imagination). It is Waterloo's description of technology that makes his story appear quaint.

Technology advances. During the 1950s, astronomers realised they could use radio waves for interplanetary communication. Radio would be cheaper, easier, and more effective than lighting beacons. (We have to assume the Martians have developed radio receivers, of course.) During the 1960s, physicists realised they could signal our presence using lasers. Indeed, we can use radio waves and laser beams to reach not just Mars, but the stars.

The search for extraterrestrial intelligence (SETI) inverts this idea. SETI listens rather than transmits. SETI is a research program based on the notion that advanced alien intelligences use radio waves or lasers to signal their existence. And to ensure receivers do not mistake the radiation as coming from a natural source, the signal carries some mathematical pattern. The 3, 4, 5 of a Pythagorean triangle, perhaps. Or the 2, 3, 5, 7, 11 … sequence of primes, maybe. Or the opening digits of pi, 3.14156 …. Or whatever.

One can identify a recency bias to this argument. The great Gauss contemplated the use of heliotropes to reflect sunlight and thus communicate with extraterrestrials. Such was the technology available to him. As we saw, astronomers turned their thoughts to radio as soon as the technology became available. As technology advanced further, they began to explore the possibilities of laser communication. At every stage, scientists have applied the most recent human technology to this problem. In the future, some other technology might seem the obvious choice. If aliens exist, maybe they communicate using techniques we cannot even understand? We have little choice here, though. We can only use the tools we have available to us.

Another aspect of "Love and a Triangle" contributes to its quaintness: the conviction that Mars is home to intelligent beings. For a time, people took this for granted. In 1877, the Italian astronomer Schiaparelli noted the presence of surface features on Mars. He called these features 'canali'. In the sense intended by Schiaparelli, the word means 'channels'. An English mistranslation into 'canals' became popular, though, and it led to confusion. *Channels*? Probably natural. *Canals*? Artificial. A mythology developed around the notion of a dying Martian civilisation. The Martians, romantics supposed, had built the canals to bring water from the polar ice caps to the parched equatorial regions. (See Fig. 3.1.)

To illustrate the extent to which the notion of a populated Mars suffused the culture of the time, consider that Stanley Waterloo did not invent the Guzman prize mentioned in his story. The prize offered 100,000 francs for the first person to communicate with extraterrestrials. Waterloo, though, got the terms of the prize wrong. In her will, Madame Guzman established the award

3 Messaging Extraterrestrial Intelligence

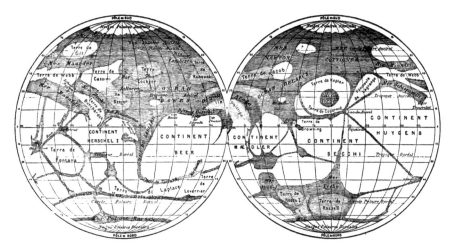

Fig. 3.1 In 1892, the French astronomer Camille Flammarion published his masterpiece "La Planete Mars et ses conditions d'habitabilite". This comprehensive summary of current knowledge about Mars discussed, among other topics, whether the red planet could sustain life. Flammarion's book contained illustrations such as this one, Fig. 31 in the original, which showed 'canals' resembling those observed a decade earlier by Schiaparelli. We now know these features do not exist, but for a while people could dream that these 'canals' demonstrated Martian technological prowess. (Credit: F0x1, CC BY-SA 4.0 International)

for communicating with a celestial body *other than Mars*. Along with the intellectuals of the day, she believed the problem of communicating with Martians too trivial a task to command such a prize!

We now know Mars is devoid of intelligent life. Mars is probably devoid of any form of life today. The red planet might well have *always* been a lifeless place. Furthermore, at the time of writing, the SETI program has failed to discover any signals from intelligence beyond our solar system. Searches for technosignatures and biosignatures have, so far, drawn a blank. We might be alone.

We *might* be alone. Perhaps the majority of scientists, though, believe we share the cosmos with other intelligent beings. For those who accept that belief, the story "Love and a Triangle" raises a profound question. Should we attempt to send a message to extraterrestrial intelligence? Should we engage in METI?

If alien intelligences exist then they are almost certainly older, and more technologically advanced, than us. If we believe those intelligences might exhibit malign intentions towards other species … well, drawing attention to ourselves would not be the smartest of moves. (Some have even postulated this as an explanation of the great silence, of the absence of signals from technological species. Perhaps the possibility of drawing attention to one's home planet terrifies those intelligent enough to contemplate such ideas. Everyone listens; no one speaks.) Even if one believes such worries are misplaced, ethical questions abound. Who, for example, should speak for Earth? What message should we as a species broadcast to the cosmos? If disagreements arise in a METI program, how should we resolve them?

We might hope that humanity would engage in a broad, rigorous, ethical analysis before we send any messages into space. But what chance of such scrutiny taking place? We live in an age in which individual billionaires have the resources to broadcast a message to the cosmos, should they so wish. People have *already* used radio telescopes such as that at Yevpatoria to send more than a dozen broadcasts to the stars (see Fig. 3.2). The star Altair heard from us in 2017. Teegarden's star received a signal in 2022. A signal from

Fig. 3.2 The Yevpatoria RT-70 radio telescope in Ukraine. The antenna has a diameter of 70 m, making the device one of the largest single-dish radio telescopes in the world. The observatory also possesses powerful transmitters, which can beam signals towards the stars. Astronomers have used Yevpatoria to send a range of messages to the cosmos. (Credit: Vyacheslav Argenberg, www.vascoplanet.com, CC BY-SA 4.0 International)

Earth will reach the star GJ 83.1 at around the time this book sees print. A further ten stars will hear from us in the next quarter of a century.

I wonder … will we get a response?

Notes and Further Reading

The Pythagorean theorem—This theorem is probably the most famous statement in mathematics, and it has applications throughout science. Although Pythagoras was possibly the first person to prove the statement (although if he did indeed have a proof it has been lost to history), Babylonian mathematicians knew of the relation more than a thousand years before Pythagoras was born. Maor (2010) traces the long history of the theorem.

Carl Friedrich Gauss, one of the greatest of all mathematicians—Gauss, the 'Prince of Mathematics', was an undoubted genius. We know from his writings that he believed in the existence of extraterrestrial life, and he thought about ways of making contact. He is supposed to have suggested planting forests on the Siberian tundra to represent the Pythagorean theorem for Martians to observe. Although Gauss may have made this suggestion, there is no definitive evidence linking him to it. We do know, however, that he contemplated using heliotropes (one of his own inventions) to send a signal to Mars. Crowe (1986) tells the story.

the largest telescope on Earth—Asimov (1976) gives an overview of the development of the telescope, including a discussion of the Leviathan, the Hooker Telescope, and others.

The search for extraterrestrial intelligence—Shuch (2011), a collection of papers published to commemorate the fiftieth anniversary of the first SETI observations by Frank Drake, contains articles giving an overview of the history of the SETI and presenting options for the future of the field.

the Italian astronomer Schiaparelli—Shindell (2023) tells the story of how people in different historical eras have thought about the red planet. His book therefore covers that exciting period during which some astronomers believed Mars might be home to intelligent life.

the Guzman prize—Oberhaus (2019) discusses some of the background to the Guzman prize.

radio telescopes such as that at Yevpatoria—Between 1999 and 2008, the Russian astronomer Aleksandr Leonidovich Zaitsev (1945–2021) used the Yevpatoria RT-70 radio telescope in Ukraine to send multiple signals towards star systems. He sent the radio signal "A Message From Earth" towards the exoplanet Gliese 581c on 9 October 2008. The signal will reach the planet in 2029. This action was controversial because Gliese 581c lies in the habitable zone of its star, with the potential for liquid water—and thus perhaps life—to exist on its surface. Although later observations concluded that the planet is tidally locked to its star, and so is unlikely to be home to advanced life forms, many have nevertheless questioned Zaitsev's wisdom of sending a signal without prior discussion and agreement; see Azua-Bustos et al. (2015), for example, for a statement on the topic signed by 28 scientists.

References

Asimov, I.: Eyes on the Universe. André Deutsch, London (1976)
Azua-Bustos, A. et al.: Regarding Messaging to Extraterrestrial Intelligence (METI)/Active Searches for Extraterrestrial Intelligence (Active SETI). https://setiathome.berkeley.edu/meti_statement_0.html (2015)
Crowe, M.J.: The Extraterrestrial Life Debate, 1750–1900. Cambridge University Press, Cambridge (1986)
Maor, E.: The Pythagorean Theorem: A 4000-Year History. Princeton University Press, Princeton (2010)
Oberhaus, D.: *Extraterrestrial Languages*. MIT University Press, Boston, MA (2019)
Shindell, M.: For the Love of Mars: A Human History of the Red Planet. University of Chicago Press, Chicago (2023)
Shuch, H.P.: Searching for Extraterrestrial Intelligence: SETI Past, Present, and Future. Springer, Berlin (2011)

4

First Contact

The Fortunate Island (Max Adeler)

Chapter I. The Island

When the good ship *Morning Star*, bound to Liverpool from New York, foundered at sea, the officers, the crew, and all of the passengers but two, escaped in the boats. Professor E. L. Baffin and his daughter, Matilda Baffin, preferred to intrust themselves to a patent India-rubber life-raft, which the Professor was carrying with him to Europe, with the hope that he should sell certain patent rights in the contrivance.

There was time enough, before the ship sank, to inflate the raft and to place upon it all of the trunks and bundles belonging to the Professor and Matilda. These were lashed firmly to the rubber cylinders, and thus Professor Baffin was encouraged to believe that he might save from destruction all of the scientific implements and apparatus which he had brought with him from the Wingohocking University to illustrate the course of lectures which he had engaged to give in England and Scotland.

Having made the luggage fast, the Professor handed Matilda down from the ship's side, and when he had tied her to one of the trunks and secured himself to another, he cut the raft adrift, and, with the occupants of the boats, sorrowfully watched the brave old *Morning Star* settle down deeper and deeper into the water; until at last, with a final plunge, she dipped beneath the surface and disappeared.

The prospect was a cheerless one for all of the party. The sea was not dangerously rough; but the captain estimated that the nearest land was at least eight hundred miles distant; and, although there were in the boats and upon the raft provisions and water enough for several days, the chance was small that a port could be made before the supplies should be exhausted. There was, moreover, almost a certainty that the boats would be swamped if they should encounter a severe storm.

The Professor, for his part, felt confident that the raft would outlive any storm; but his shipmates regarded his confidence in it as an indication of partial insanity.

The captain rested his expectations of getting ashore chiefly upon the fact that they were in the line of greatest travel across the Atlantic, so that they might reasonably look to meet, within a day or two, with a vessel of some kind which would rescue them.

As the night came on, it was agreed that the boats and the raft should keep together, and the captain had provided a lantern, which was swung, lighted, aloft upon an oar, so that the position of his boat could be determined. The Professor, with his raft under sail, steered along in the wake of the boats for several hours, Matilda, meanwhile, sleeping calmly, after the exciting and exhausting labours of the day, upon a couple of trunks.

As the night wore on, a brisk wind sprang up, and shortly afterward the light upon the captain's boat for some reason disappeared. The Professor was somewhat perplexed when he missed it, but he concluded that the safest plan would be to steer about upon the course he had hitherto held, and then to communicate with the boats if they should be within sight in the morning.

The wind increased in force about midnight, and the raft rolled and pitched in such a manner that the Professor's faith in it really lost some of its force. Several times huge waves swept over it, drenching the Professor and his daughter, and filling them with grave apprehensions of the result if the storm should become more violent.

Even amid the peril, however, Professor Baffin could not but admire the heroic courage and composure of Matilda, who sat upon her trunk, wet and shivering with cold, without showing a sign of fear, but trying to encourage her father with words of hope and cheer.

When the dawn came, dim and gray, the gale abated its force, and although the sea continued rough, the raft rode the waves more buoyantly and easily. Producing some matches from his waterproof box, the Professor lighted the kerosene-lamp in the tiny stove which was in one of the boxes; and then Matilda, with water from the barrel, began to try to make some coffee. The attempt seemed to promise to be successful, and while the process was going on, the Professor looked about for the boats. They could not be seen. The Professor took out his glass and swept the horizon. In vain; the boats had disappeared completely; but the Professor saw something else that attracted his attention, and made his heart for a moment stop beating.

Right ahead, not distinctly outlined, but visible in a misty sort of way, he thought he discerned land!

At first he could not believe the evidence of his sight. The captain, an expert navigator, had assured him that they were eight hundred miles from any shore. But this certainly looked to the Professor very much like land. He examined it through his glass. Even then the view was not clear enough to remove all doubts, but it strengthened his conviction; and when Matilda looked she said she knew it was land. She could trace the outline of a range of hills.

"Tilly," said the Professor, "we are saved! It *is* the land, and the raft is drifting us directly towards it. We cannot be sufficiently thankful, my child, for this great mercy! Who would have expected it? Taken altogether, it is the most extraordinary circumstance within my recollection."

"Captain Duffer must have made a miscalculation," said Tilly. "The ship must have been off of her course when she sprang a leak."

"It is incomprehensible how so old a sailor could have made such a blunder," replied the Professor. "But there the land is; I can see it now distinctly. It looks to me like a very large island."

"Are you going ashore at once, pa?"

"Certainly, dear; that is, if we can make a landing through the breakers."

"Suppose there are cannibals on it, pa? It would be horrid to have them eat us!"

"They would have to fatten us first, darling; and that would give us an opportunity to study their habits. It would be extremely interesting!"

"But the study would be of no use if they should eat us!"

"All knowledge is useful, Tilly; I could write out the results of our observations, and probably set them adrift in a bottle!"

"It is such a dreadful death!"

"Try to look at it philosophically! There is really nothing more unpleasant about the idea of being digested than there is about the thought of being buried."

"O, pa!"

"No, my child! It is merely a sentiment. If I shall be eaten, and we have volition after death, I am determined to know how I agreed with the man who had me for dinner! Tilly, I have a notion that you would eat tender!"

"Pa, you are simply awful!"

"To me, indeed, there is something inspiring in the thought that my physical substance, when I have done with it, should nourish the vitality of another being. I don't like to think that I may be wasted."

"You seem as if you rather hoped we should find savage cannibals upon the island!"

"No, Tilly; I hope we shall not. I believe we shall not. Man-eaters are rarely found in this latitude. My impression is that the island is not inhabited at all. Probably it is of recent volcanic origin. If so, we may have a chance to examine a newly-formed crater. I have longed to do so for years."

"We might as well be eaten as to be blown up and burned up by a volcano," said Matilda.

"It would be a grand thing, though, to be permitted to observe, without interruption, the operation of one of the mightiest forces of nature! I could make a magnificent report to the Philosophical Society about it; that is, if we should ever get home again."

"For my part," said Matilda, "I hope it contains neither cannibals nor volcanoes; I hope it is simply a charming island without a man or a beast upon it."

"Something like Robinson Crusoe's, for example! I have often thought I should like to undergo his experiences. It must be, to an inquiring mind, exceedingly instructive to observe in what manner a civilized man, thrown absolutely upon his own resources, contrives to conduct his existence. I could probably enrich my lecture upon Sociology if we should be compelled to remain upon the island for a year or two."

"But we should starve to death in that time!"

"So we should; unless, indeed, the island produces fruits of some kind from its soil. I think it does. It seems to be covered with trees, Tilly, doesn't it?"

"Yes," said Matilda, looking through the glass. "It is a mass of verdure. It is perfectly beautiful. I believe I see something that looks like a building, too."

"Impossible! you see a peculiar rock formation, no doubt; I shan't be surprised if there is enough in the geological formation of the island to engage my attention so long as we remain."

"But what am I to do, meantime?"

"You? Oh, you can label my specimens and keep the journal; and maybe you might hunt around for fossils a little yourself."

The raft rapidly moved toward the shore, and the eyes of both of the voyagers were turned toward it inquiringly and eagerly. Who could tell how long the island might be their home, and what strange adventures might befall them there?

"The wind is blowing right on shore, Tilly," said the Professor. "I will steer straight ahead, and I shouldn't wonder if we could shoot the breakers safely.

Isn't that a sand-beach right in front there?" inquired the Professor, elevating his nose a little, to get his spectacles in focus. "It looks like one."

"Yes, it is," replied Matilda, looking through her glass.

"First-rate! Couldn't have been better. There, we will drive right in. Tilly, hoist my umbrella, so as to give her more sail!"

The raft fairly danced across the waves under the increased pressure, and in a moment or two it was rolling in the swell just outside of the line of white breakers. Before the Professor had time to think what he should do to avoid the shock, a huge wave uplifted the raft and ran it high upon the beach with such violence as to compel the Professor to turn a somersault over a trunk. He recovered himself at once, and replacing his spectacles he proceeded, with the assistance of Matilda, to pull the raft up beyond the reach of the waves.

Then, wet and draggled, with sand on his coat, and his hat knocked completely out of shape, he stood rubbing his chin with his hand, and thoughtfully observing the breakers.

"Extraordinary force, Tilly, that of the ocean surf,—clear waste, too, apparently. If we stay here long enough, I must try to find out the secret of its motion."

"Hadn't we better put on some dry clothing first?" suggested Miss Baffin, "and examine the surf afterwards? For my part I have had enough of it."

"Certainly! Have you the keys of the trunks? Everything soaking wet, most likely."

When the trunks were unfastened, the Professor was delighted to find that the contents were perfectly dry. Selecting some clothing for himself, he went behind a huge rock and proceeded to dress. Matilda, after looking carefully about, retreated to a group of trees, and beneath their shelter made her toilette.

"Isn't this a magnificent place?" said the Professor, when Matilda, nicely dressed, came out to where he was standing by the raft.

"Perfectly lovely."

"Noble trees, rich grass, millions of wild flowers, birds twittering above us, a matchless sky, a bracing air, and—why, halloa! there's a stream of running water! We must have a drink of that, the very first thing. Delicious, isn't it?" asked the Professor, when Miss Baffin, after drinking, returned the cup to him.

"It is nectar."

"I tell you what, Tilly, I am not sure that it wouldn't be a good thing to be compelled to live here for two or three years. The vegetation shows that we are in a temperate latitude, and I know I can find or raise enough to eat in such a place as this."

"Why, pa, look there!"

"Where?"

"Over there. Don't you see that castle?"

"Castle? No! What! Why, yes, it is! Bless my soul, Tilly, the place is inhabited!"

"Who would have thought of finding a building like that on an island in mid-ocean?"

"It is the most extraordinary circumstance, taking it altogether, that ever came under my observation," said the Professor, looking towards the distant edifice. "So far as I can make out, it is a castle of an early period."

"Mediæval?"

"Well, not later than the seventh or eighth century, at the farthest. Tilly, I feel as if something remarkable was going to happen."

"Pa, you frighten me!"

"No, I mean something that will be extraordinarily interesting. I know it. The voice of instinct tells me so. Have you your journal with you?"

"It is in the trunk."

"Get it and your lead-pencils. We will drag the baggage further up from the water, and then we will push towards the castle. I am going to know the date of that structure before I sleep tonight."

"There can hardly be any danger, I suppose?" suggested Miss Baffin, rather timidly.

"Oh, no, of course not; I have my revolver with me. Let me see; where is it? Ah, here. And the cartridges are waterproof. I think I will put a few things in a valise, also. We might find the castle empty, and have to depend upon ourselves for supper."

The Professor then let the air out of the raft, and folded the flattened cylinders together.

When the valise was ready, the Professor grasped it, shouldered his umbrella, and said, "Now, come, darling, and we will find out what all this means."

The pair started along a broad path which ran by the side of the stream, following the course of the brook, and winding in and out among trees of huge girth and gigantic height. Birds of familiar species flitted from branch to branch before them, as if to lead them on their way; now and then a brown rabbit, after eyeing them for a moment with quivering nostrils, beat a quick tattoo upon the ground with his hind legs, then threw up his tail and whisked into the shrubbery. Gray squirrels scrambled around the trunks of the trees to look at them, and now and then a screaming, blue-crested kingfisher ceased his complaining while he plunged into one of the pools of the rivulet, and emerged with a trout in his talons.

It was an enchanting scene; and Miss Baffin enjoyed it thoroughly as she stepped blithely by the side of her father, who seemed to find especial pleasure in discovering that the herbage, the trees, the rocks, and all the other natural objects, were precisely like those with which he had been familiar at home.

After following the path for some time, the pair came to a place where the brook widened into a great pool, through which the water went sluggishly, bearing upon its surface bubbles and froth, which told how it had been tossed and broken by rapid descents over the rocks in some narrow channel above. Here the Professor stopped to observe an uncommonly large and green bull-frog, which sat upon a slimy stone a few yards away, looking solemnly at him.

During the pause, they were startled to hear a voice saying to them,—

"Good morrow, gentle friends."

Matilda uttered a partly-suppressed scream, and even the Professor jumped backward a foot or two, in astonishment.

Looking toward the place from which the voice came, they saw an old man with gray hair and beard lifting a large stone pitcher, which he had been filling from the pool. He was dressed in a long and rather loose robe, which reached from his shoulders to his feet, and which was gathered about his waist with a knotted cord. This was his entire costume, for his feet were bare, and he wore no hat to hide the rich masses of hair which fell to his shoulders. As he offered his salutation, he raised his pitcher until he stood upright, and then he looked at the Professor and Miss Baffin with a pleasant smile, in which there were traces of curiosity.

"Good afternoon," returned the Professor, after a moment's hesitation; "how are you?"

"Are you not strangers in this land?" asked the old man.

"Well, yes," said the Professor, briskly, with a manifest purpose to be sociable; "we have just come ashore down here on the beach. Shipwrecked, in fact. This is my daughter. Let me introduce you. My child, allow me to make you acquainted with—with—beg pardon, but I think you did not mention your name."

"I am known as Father Anselm."

"Ah, indeed! Matilda, this is Father Anselm. A clergyman, I suppose?"

"I am a hermit; my cell is close at hand. You will be welcome there if you will visit it."

"A hermit! Living in a cell! Well, this *is* surprising! We shall be only too happy to visit you, if you will permit us. Delightful, isn't it, dear? We will obtain some valuable information from the old gentleman."

The Hermit, with the pitcher poised upon his shoulder, led the way, and he was closely followed by the Professor and by Matilda, who regarded the proceeding rather with nervous apprehension. The Hermit's cell was a huge cave, excavated from the side of a hill. The floor was covered with sprigs of fragrant evergreens. A small table stood upon one side of the apartment; beside it was a rough bench, which was the only seat in the room. A crucifix, a candle, a skull, an hour-glass, and a few simple utensils were the only other articles to be seen.

The Hermit brought forward the bench for his visitors to sit upon, and then, procuring a cup, he offered each a drink of water.

The Professor, hugging one knee with interlocked fingers, seemed anxious to open a conversation.

"Pardon me, sir, but do I understand that you are a clergyman; that is to say, some sort of a teacher of religion?"

"I belong to a religious order. I am a recluse."

"Roman Catholic, I presume?" said the Professor, glancing at the crucifix.

"Your meaning is not wholly clear to me," replied the Hermit.

"What are your views? Do you lean to Calvinism, or do you think the Arminians, upon the whole, have the best of the argument?"

"The gentleman does not understand you, pa," said Miss Baffin.

"Never mind, then; we will not press it. But I should like very much if you would tell us something about this place; this country around here," said the Professor, waving his hand towards the door.

"Let me ask first of the misadventure which cast you unwillingly upon our shores?" said the Hermit.

"Well, you see, I sailed from New York on the twenty-third of last month, with my daughter here, to fulfil an engagement to deliver a course of lectures in England."

"In England!" exclaimed the Hermit, with an appearance of eager interest.

"Yes, in England. I am a professor, you know, in an American university. When we were about half way across, the ship sprang a leak, from some cause now unknown. My daughter and I got off with our baggage upon a life-raft, which I most fortunately had with me. The rest of the passengers and the crew escaped in the boats. I became separated from them, and drifted here. That is the whole story."

"I comprehend only a part of what you say," replied the Hermit. "But it is enough that you have suffered; I give you hearty welcome."

"Thank you. And now tell me where I am."

"You spoke of England a moment ago," said the Hermit. "Let me begin with it. Hundreds of years ago, in the time of King Arthur, of noble fame, it happened, by some means even yet not revealed to us, that a vast portion of that island separated from the rest, and drifted far out upon the ocean. It carried with it hundreds of people—noble, and gentle, and humble. This is that country."

"In-*deed*!" exclaimed the Professor. "This? This island that we are on? Amazing!"

"It is true," responded the Hermit.

"Why, Tilly, do you hear that? This is the lost Atlantis! We have been driven ashore on the far-famed Fortunate Island! Wonderful, isn't it? Taking every thing into consideration, I must say this certainly is the most extraordinary circumstance I ever encountered!"

"Nobody among us has ever heard anything from England or of it, excepting through tradition. No ship comes to our shores, and those of us who have built boats and gone away in search of adventure have never come back. Sometimes I think the island has not ended its wanderings, but is still floating about; but we cannot tell."

"But, my dear sir," said the Professor, "you can take your latitude and longitude at any time, can't you?"

"Take what?"

"Your latitude and longitude! Find out exactly in what part of the world you are?"

"I never heard that such a thing was done. None of our people have that kind of learning."

"Well, but you have schools and colleges, and you acquire knowledge, don't you?"

"We have a few schools; but only the low-born children attend them, and they are taught only what their fathers learned. We do not try to know more. We reverence the past. It is a matter of pride among us to preserve the habits, the manners, the ideas, the social state which our fore-fathers had when they were sundered from their nation."

"You live here pretty much as King Arthur and his subjects lived?"

"Yes. We have our chivalry; our knight errants; our tournaments; our castles—everything just as it was in the old time."

"My dear," said the Professor to Miss Baffin, "the wildest imagination could have conceived nothing like this. We shall be afforded an opportunity to study the middle ages on the spot."

"Sometimes," said the Hermit, gravely, "I have secret doubts whether our way is the best, whether in England and the rest of the world men may not have learned while we have remained ignorant; but I cannot tell. And no one would be willing to change if we could know the truth."

"My friend," said the Professor, with a look of compassion, "the world has gone far, far ahead of King Arthur's time! It has almost forgotten that there ever was such a time. You would hardly believe me, at any rate you would not understand me, if I should tell you of the present state of things in the world.

But if I stay here I will try to enlighten you gradually. I feel as if I had been sent here as a missionary for that very purpose."

"Do you come from England?"

"Oh, no! I was going thither. I came from the United States. You never heard of them, of course. They are a land right across the ocean from England, about three thousand miles."

"Discovered by a man named Columbus," said Miss Baffin.

"Your dress is an odd one," continued the Hermit. "Are you a fighting man?"

"A fighting man! Oh, no, of course not. I'm a Professor."

"Then this is not a weapon that you carry."

"Bless my soul, my dear sir! Why, this is an umbrella! Tilly, we have to deal with a very primitive condition of things here. It is both entertaining and instructive."

"What is it for?"

"I will show you. Suppose it begins to rain, I untie this string and open the umbrella, *so!* Now don't be alarmed! It is perfectly harmless, I assure you!"

The holy man had retreated suddenly into the furthest recess of the cell.

"While it rains I hold it in this manner. When it clears, I shut it up, *thus*, and put it under my arm."

"Wonderful! wonderful!" exclaimed the Hermit. "I thought it was an implement of war. The world beyond us evidently has surpassed us."

"This is nothing to the things I will show you," said the Professor. "I see you have an hour-glass here. Is this the only way you have of recording time?"

"We have the sun."

"No clocks or watches?"

"I do not know what they are."

"Tilly, show him your watch. This is the machine with which we tell time."

"Alive, is it?" asked the Hermit.

The Professor explained the mechanism to him in detail.

"You are indeed a learned man," said the recluse. "But I have forgotten a part of my duty. Will you not take some food?"

"Well," said the Professor, "if you have anything about in the form of a lunch, I think I could dispose of it."

"I am awfully hungry," said Miss Baffin.

The Hermit produced a piece of meat, and hanging it upon a turnspit he gathered a few sticks and placed them beneath it. The Professor watched him closely; and when the holy man took in his hands a flint and steel with which to ignite the wood, the Professor exclaimed,—

"One moment! Let me start that fire for you?"

Taking from his pocket an old newspaper, he put it beneath the sticks; then from his match-box he took a match, and striking it there was a blaze in a moment.

The Hermit crossed himself and muttered a prayer at this performance.

"No cause for alarm, I assure you," said the Professor.

"You must be a wizard," said the Hermit.

"No; I did that with what we call a match; like this one. There is stuff on the end which catches fire when you rub it," and the Professor again ignited a match.

"I never could have dreamed that such a thing could be," exclaimed the recluse. "You will be regarded by our people as the most marvellous magician that ever lived."

The Professor laughed.

"Oh," said he, "I will let them know it is not magic. We must clear all that nonsense away. Tilly, I feel that duty points me clearly to the task of delivering a course of lectures upon this island."

During the repast, the Hermit, looking timidly at Professor Baffin, said,—

"Would it seem discourteous if I should ask you another question?"

"Certainly not. I shall be glad to give you any information you may want."

"What, then," inquired the Hermit, "is the reason why you protect your eyes with glass windows?"

"These," said the Professor, removing his spectacles, "are intended to improve the sight. I cannot see well without them. With them I have perfect vision. Tilly, make a memorandum in the journal that my first lecture shall be upon Optics."

"Pa, I wish we could learn something about the castle we saw," observed Miss Baffin.

"Oh, yes; by the way, Father Anselm," said the Professor, "we observed an old-fashioned castle over yonder, as we came here. Can you tell me anything about it?"

"The castle," replied the Hermit, "is the home and the stronghold of Sir Bors, Baron of Lonazep. He is a great and powerful noble, much feared in this country."

"Any family?" inquired the Professor.

"He has a gallant son, Sir Dinadan, as brave a knight as ever levelled lance, and a beautiful daughter, Ysolt. Both are unmarried; but the fair Ysolt fondly loves Sir Bleoberis, to whom, however, the Baron will not suffer her to be wedded, because Sir Bleoberis, though bold and skilful, has little wealth."

"Human nature, you observe, my child, is the same everywhere. We have heard of something like this at home," remarked the Professor to his daughter.

"Ysolt is loved also by another knight, Sir Dagonet. He has great riches, and is very powerful; but he is a bad and dangerous man, and the Baron will not consent to give him Ysolt to wife. These matters cause much strife and much unhappiness."

"It's the same way with us," observed the Professor; "I have known lots of such cases."

"I hope we shall stay here long enough to see how it all turns out," said Miss Baffin.

"Of course," replied the Professor. "You hated the island when you thought it might promote the interests of science. But some lovers' nonsense would keep you here willingly for life. Just like a woman."

"The King," said the Hermit, "has espoused the cause of Sir Bleoberis, and we hope he may win the lady for the knight whom she loves."

"The King, eh? Then you have a monarchical government?"

"We have eleven kings upon this island."

"All reigning?"

"Yes."

"How many people are there in the whole island?"

"No one knows, exactly. One hundred thousand, possibly."

"Not ten thousand men apiece for the kings! Humph! In my country we have a million men in one town, and nobody but a common man to rule them."

"Incredible!"

"And what is the name of your particular king,—the one who is lord of this part of the country?"

"King Brandegore; a wise, and good, and valiant monarch."

"Tilly," said the Professor, "you might as well jot that down. Eleven kings on the island, and King Brandegore running this part of the government. I must get acquainted with him."

When the meal was finished the Professor said to the recluse,—

"Do you allow smoking?"

"Smoking!"

"Pray excuse me! I forgot. If you will permit me, I will introduce you to another of the practices of modern civilization."

Then the Professor lighted a cigar, and, sitting on the bench in a comfortable position, with his back against the wall of the cave, he began to puff out whiffs of smoke.

The Hermit, with a look of alarm, was about to ask for an explanation of the performance, when loud cries were heard outside of the cave mingled with frightened exclamations from a woman.

The occupants of the cavern started to their feet, just as a beautiful girl, dressed in a quaint but charming costume, ran into the doorway in such haste that she dashed plump up against the Professor, who caught her in his arms.

For a moment she was startled at seeing two strangers in a place where she had thought to encounter none but the Hermit; but her dread of her pursuer overcame her diffidence, and, clinging to the Professor, she exclaimed,—

"Oh, save me! save me!"

"Certainly I will," said the Professor, soothingly, as his arm tightened its clasp about her waist. "What's the matter? Don't be afraid, my child. Who is pursuing you?"

The Professor was not displeased at the situation in which he found himself. The damsel was fair to see, and the head which rested, in what seemed to him sweet confidence, upon his shoulder, was crowned with golden hair of matchless beauty. Even amid the intense excitement of the moment the reflection flashed through the Professor's mind that he was a widower, and that Matilda had always expressed a willingness to try to love a stepmother.

"My father! The Baron! He threatens to kill me," sobbed the maiden, and then, tearing herself away from the Professor in a manner which struck him as being, to say the least, inconsiderate, she flew to Father Anselm and said, "You, holy father, will save me."

"I will try, my daughter; I will try," replied the Hermit. And then, turning to the Professor he said, "It is Ysolt."

"Ah!" said the Professor, "the Baron's daughter. May I ask you, miss, what the old gentleman is so excited about? It is not one of the customs here for indignant parents to chase their children around the country, is it?"

"I had gone from the castle," said the damsel, partly to the Hermit and partly to Professor Baffin, "to meet Sir Bleoberis at the trysting-place. My father was watching me, and as I neared the spot he rushed toward me with a drawn sword, threatening to kill me."

"It is an outrageous shame!" exclaimed the Professor, sympathetically.

"I eluded him," continued the sobbing girl, "and flew towards this place. When he saw me at last he gave chase. I am afraid he will slay me when he comes."

"I think, perhaps, I may be able to reason with this person when he arrives," said the Professor, rubbing his chin and looking at the hermit over the top of his spectacles. "The Baron ought to be ashamed of himself to go on in this manner! Tilly, wipe the poor creature's eyes with your handkerchief. There now, dear, cheer up."

Just then the Baron rushed into the cell, with his eyes flaming, and his breath coming short and fast.

He was a large man, with a handsome face, thick covered with beard. He was dressed in doublet, trunks and hose, and over one shoulder a mantle hung gracefully. His sword was in its sheath, and it was manifest that he had repented of his murderous purpose.

"Where is that faithless girl?" he demanded in a voice of thunder.

Ysolt had hidden behind Matilda Baffin.

"Say, priest, where have you secreted her?"

"One moment!" said the Professor, stepping forward. "May I, without appearing impertinent, offer a suggestion?"

"Out, varlet!" exclaimed the Baron, pushing him aside. "Tell me, Hermit, where is Ysolt?"

The Professor was actually pale with indignation. Pushing himself in front of the Baron, and brandishing his umbrella in a determined way he said:

"Old man, I want you to understand that you have to deal with a free and independent American citizen! What do you mean by 'varlet'? I hurl the opprobrious word back into your teeth, sir! I am not going to put up with such conduct, I'd like you to know!"

The Baron for the first time perceived what manner of man the Professor was, and he paused for a moment amid his rage to eye the stranger with astonishment.

"Why do you want to hurt the young woman? Is this any way for an affectionate father to behave to his own offspring? Allow me to say, sir, that I'll be hanged if I think it is! If you don't want her to marry Sir What's-his-name, don't let her; but it strikes me that charging around the country after her, and threatening to kill her, is an evidence that you don't understand the first principles of domestic discipline!"

"What do you mean? Who are you? What are you doing here?" demanded the Baron, fiercely, recovering his self-possession.

"I am Professor E. L. Baffin, of Wingohocking University; and I mean to try to persuade you to treat your daughter more gently," said the Professor, cooling as he remembered that the Baron had a father's authority.

"You have a weapon. I will fight you," said the Baron, drawing his sword.

The Professor put his cigar in his mouth, and opened his umbrella suddenly in the Baron's face.

The Baron retreated a distance of twenty feet and looked scared.

"Come," said the Professor, closing his umbrella and smiling, "I am not a fighting man. We will not quarrel. Let us talk the matter over calmly."

But the Baron, mortified because of the alarm that he had manifested, rushed savagely at the Professor, and would have felled him to the earth had not Matilda sprung forward and placed herself, shrieking, between the Baron and her father.

At this precise juncture, also, a young man entered the cell, and, seeing the Baron apparently about to strike a woman, seized his sword-arm and held it. The Baron turned sharply about. Recognizing the youth as his son, he simply looked at him angrily, and then, while Miss Baffin clung to the Professor, the Baron seized Ysolt by the arm and led her weeping away.

The Professor, after freeing himself from Miss Baffin's embrace, extended his hand to the youth, and said,—

"I have not the honour of knowing you, sir, but you have behaved handsomely. Permit me to inquire your name?"

"Sir Dinadan; the son of the Baron," said the youth, taking hold of the Professor's hand, as if he were somewhat uncertain what he had better do with it.

"No last name?" asked the Professor.

"That is all. And you are?—"

"I am Everett L. Baffin, a Professor in the Wingohocking University. I was cast ashore down here with my daughter. Tilly, let me introduce to you Sir Dinadan."

Sir Dinadan coloured, and dropping upon his knee he seized Miss Baffin's hand and kissed it. Rising, he said:

"What, Sir Baffin, is the name of the sweet lady?"

"Matilda."

"How lovely!" exclaimed Sir Dinadan.

"It is abbreviated sometimes to Tilly, by her friends."

"It is too beautiful," said the youth, gazing at Miss Baffin with unconcealed admiration. "I trust, Sir Baffin, I may be able to serve in some manner you and the Lady Tilly."

"Professor Baffin, my dear sir; not Sir Baffin. Permit me to offer you my card."

Sir Dinadan took the card, and seemed perplexed as to its meaning. He turned it over and over in a despairing sort of way in his fingers.

"If you will read it," said the Professor, "you will find my name upon it."

"But, Sir Baffin, I cannot read."

"Can't read!" exclaimed the Professor, in amazement. "You don't mean to say that you have never learned to read!"

"High-born people," replied Sir Dinadan, with an air of indifference, "care nothing for learning. We leave that to the monks."

"This," said the Professor to Miss Baffin, "is one of the most extraordinary circumstances that has yet come under my observation. Tilly, mention in your journal that the members of the upper classes are wholly illiterate."

"As the Lady Tilly is a stranger here," said Sir Dinadan, "I would be glad to have her walk with me to the brow of the hill. I will show her our beautiful park."

"That would be splendid!" said Miss Baffin. "May I go, pa?"

"Well, I don't know," said the Professor, with hesitation, and looking inquiringly at the Hermit. As that individual appeared to regard the proposition with no such feeling of alarm as would indicate a breach of ordinary social custom, the Professor continued, "Yes, dear, but be sure not to go beyond ear-shot."

Sir Dinadan, smiling, led Miss Baffin away, and the Professor sat down to finish his cigar and to have some further conversation with the Hermit. Before he had time to begin, two other visitors arrived. Both were young men, gaily dressed in rich costume. One of them, whom the recluse greeted as Sir Bleoberis, had a tall slender figure and an exceedingly handsome countenance, which was adorned with a moustache and pointed beard. His companion, Sir Agravaine, was smaller, less comely, and if his face was an index of his mind, by no means so intelligent.

After being presented to the Professor, whom they regarded with not a little curiosity, Sir Bleoberis said:

"Holy father, the fair Ysolt was here and was taken away by the Baron, was she not?"

"Yes!"

"Alas!" said the Knight, "I see no hope. Whilst I am poor, the Baron will never relent."

"Never!" chimed in Sir Agravaine.

"Is your poverty the only objection he has to you?" asked the Professor.

"Yes."

"Well," replied the Professor, "I can understand a father's feelings in such a case. It seems hard upon a young man, but naturally he wants his daughter to be comfortable. Is there nothing you can turn your hand to to improve your fortunes?"

"We might rob somebody," said Sir Agravaine, with a reflective air.

"Rob somebody!" exclaimed the Professor, "That is simply atrocious! Can't you go to work; go into business, start a factory, speculate in stocks, or something of that kind?"

"Persons of my degree never work," said Sir Bleoberis.

The Professor sighed, "Ah! I forgot. We must think of something else. Let me see; young man, I think I can help you a little, perhaps. You agree to accept some information from me and I believe I can make your fortune."

"Do you propose," asked Sir Agravaine, "to drug the Baron, or to enchant him so that he will change his mind? I have often tried love-philters with ladies whose hands I sought, but they always failed."

"Nonsense!" exclaimed the Professor. "I don't operate with such trumpery as that. You agree to help me, and we'll give this island such a stirring up as will revolutionize it."

The Professor then proceeded to explain in detail the nature and operation of some of the scientific apparatus which he had with him in his trunk; and the Knight and the Hermit listened with open-eyed amazement while he told them of the telegraph, the telephone, the phonograph, the photograph, and other modern inventions.

Whilst the Professor waxed eloquent, Sir Dinadan and Miss Baffin strolled slowly back towards the cave.

Sir Dinadan had improved the opportunity to offer Miss Baffin his hand, rather abruptly.

"But you can try to love me," he pleaded, as she, with much embarrassment but with gentleness, resisted his importunity.

"I can try, Sir Dinadan," she said, blushing, "but really I have known you only a few moments. It is impossible for me now to have any affection for you."

"Will tomorrow be time enough?"

"No, no! I must have a much longer time than that."

"I will fight for you. We will get up a tournament and you will see how I can unhorse the bravest knights. If I knock over ten, will that make any difference in your feelings?"

"Not the slightest!"

"Fifteen?"

"You do not understand. It is not the custom in our country to press a suit upon a lady by poking people off of a horse."

"Perhaps I ought to fight your father? Will Sir Baffin break a lance with me to decide if I shall have you?"

"My father does not fight."

"Does not fight! Certainly you don't mean that?"

"He is the Vice-President of the Universal Peace Society."

"The WHAT?" asked Sir Dinadan, in amazement.

"Of the Peace Society; a society which opposes fighting of every kind, under any circumstances."

It was a moment or two before Sir Dinadan could get his breath. Then he said—

"But—but then, Lady Tilly, what—what do men in your country do with themselves?"

Miss Baffin laughed and endeavoured to explain to him the modern methods of existence.

"I never could have believed such a thing from other lips," said Sir Dinadan. "It is marvellous. But tell me, how do lovers woo in your land?"

"Really, Sir Dinadan," replied Miss Baffin, blushing, "I have had no experience worth speaking of in such matters. I suppose, perhaps, they show a lady that they love her, and then wait until she can make up her mind."

"I will wait, then, as long as you wish."

"But," said Miss Baffin, shyly, although plainly she was beginning to feel a genuine interest in the proceeding, "your father and your mother may not think as you do; and then, I shall not want to stay upon this island if I can get away."

"My mother always consents to anything I wish, and the Baron never dares to oppose what she wants. And if you go back to your own country, I will go with you, whether you accept me or not."

Miss Baffin smiled. Sir Dinadan was in earnest, at any rate. She could not help thinking of the sensation that would be created in Wingohocking if she should walk up the fashionable street of the town some afternoon with Sir Dinadan in his parti-coloured dress of doublet and stockings, and jaunty feathered cap, and sword, while his long yellow hair dangled about his shoulders.

While Sir Dinadan was protesting that he should love her for ever and for ever, they came back again to the Hermit's cell, and then Sir Dinadan, greeting Sir Bleoberis and Sir Agravaine, presented Miss Baffin to them.

Sir Bleoberis was courteous but somewhat indifferent; Sir Agravaine, upon the contrary, appeared to be deeply impressed with Miss Baffin's beauty. After gazing at her steadily for a few moments, he approached her, and while the other members of the company engaged in conversation, he said,—

"Fair lady, you are not married?"

"No, sir," replied Miss Baffin, with some indignation.

"Permit me, then, to offer you my hand."

"What!" exclaimed Miss Baffin, becoming angry.

"I love you. Will you be mine?" said Sir Agravaine, falling upon one knee and trying to take her hand.

Miss Baffin boxed his ear with a degree of violence.

Rising with a rueful countenance, he said,—

"Am I to understand, then, that you decline the offer?"

Miss Baffin, without replying, walked away from him and joined her father.

Sir Dinadan was asking the Hermit for a few simples with which to relieve the suffering of his noble mother.

"I judge, from what you say," remarked the Professor, "that the Baroness is afflicted with lumbago. The Hermit's remedies, I fear, will be ineffectual. Permit me to recommend you to iron her noble back, and to apply a porous plaster."

Sir Dinadan wished to have the process more clearly explained. The Professor unfolded the matter in detail, and said,—

"I have some plasters in my trunk, down there upon the beach."

"Then you are a leech?" asked Sir Dinadan.

"Matilda, my child," remarked the Professor, "observe that word 'leech' used by Sir Dinadan! How very interesting it is! Not exactly a leech, Sir Dinadan; but it is my habit to try to know a little of everything."

"Can you cast a lover's horoscope?" asked Sir Agravaine, looking at Matilda.

"Young man," said the Professor, sternly, "there is no such foolery as a horoscope; and as for love, you had better let it alone until you have more wit and a heavier purse."

"I wish you and the Lady Tilly to come with me to the castle," remarked Sir Dinadan. "My father will welcome you heartily if you can medicine the sickness of my mother; and she will be eager to receive your fair daughter."

"I will go, of course," replied the Professor; "you are very kind. Tilly, we had better accept, I think?"

Miss Baffin was willing to leave the matter wholly in the hands of her father.

After requesting Sir Dinadan to have his luggage brought up from the beach, the Professor bade adieu to the Hermit, and then turning to Sir Bleoberis, who stood with a disconsolate air by the fire, he said:

"I will see you again about your affair; and meantime you may depend upon my using my influence with the Baron to remove his prejudices. I will dance at your wedding yet; that is, figuratively speaking, of course; for, as a precise matter of fact, I do not know how to dance."

As the Professor and Sir Dinadan and Miss Baffin left the cell, Sir Agravaine approached the lady and whispered:

"Did I understand you to say you don't love me?"

Miss Baffin twitched the skirt of her gown to one side in a scornful way, and passed on without replying.

"Women," sighed Sir Agravaine, as he looked mournfully after her, "are *so* incomprehensible. I wish I knew what she meant."

Chapter II. The Castle of Baron Bors

As Sir Dinadan led the Professor and Miss Baffin along the lovely path which went winding through the woods toward the castle, the Professor lighted another cigar, and in response to Sir Dinadan, he entered upon an explanation of the nature of tobacco, the methods and extent of its use, and its effect upon the human system.

"The Lady Tilly, of course she smokes sometimes, also?" asked Sir Dinadan.

"Oh, no," replied Miss Baffin, "ladies in my country never do."

"Of course not," added the Professor.

"And yet, if it is so pleasing and so beneficial as you say," responded the youth, "why should not ladies attempt it?"

The Professor really could not say; Sir Dinadan was pressing him almost too closely. He compromised further discussion by yielding promptly, although with a melancholy reflection that his store of cigars was small, to a request to teach Sir Dinadan, at the earliest opportunity, to smoke.

As they neared the castle, the Professor's attention was absorbed in observing the details of the structure. It was a massive edifice of stone, having severe outlines and no ornamentation worthy of the name, but presenting, from the very grandeur of its proportions, an impressive and not unpleasing appearance. It was surrounded by a wide fosse filled with water; and the Professor was delighted to observe, as they drew near, that the entrance was protected with a portcullis and a drawbridge. The bridge was drawn up, and the iron portcullis, made of bars of huge size, was closed.

"Magnificent, isn't it, Tilly?" exclaimed the Professor, gleefully. "It is probably the most perfect specimen of early English architecture now upon earth. Most fortunately I have in my trunks a photographic apparatus with which to obtain a picture of it."

Sir Dinadan seized a curved horn which hung upon the branch of a tree, and blew a blast loud and long upon it.

The Professor regarded the performance with intense interest and not a little enthusiasm.

The warder of the castle appeared at the grating, and, perceiving Sir Dinadan, saluted him; then lowering the drawbridge and lifting the portcullis, which ascended with many hideous creaks and groans from the rusty iron, Sir Dinadan and his companions entered.

Leaving the Professor and Miss Baffin comfortably seated in a great hall, the walls of which were adorned with curious tapestries dark with age, with swords and axes and trophies of the chase, Sir Dinadan went in search of the Baron.

"Little did we think, Tilly," said the Professor, looking around, "when we left New York four weeks ago—it seems more like four years—that we should find ourselves, within a month, in such a place as this."

"I can hardly believe it yet," responded Miss Baffin.

"It does seem like a dream. And yet we are certainly wide awake, and we are in the hall of a real castle, waiting for real people to come to us."

"Sir Dinadan seems very real, too," said Miss Baffin, timidly.

"Very! There can be no doubt about it."

"And he behaves like a real young man, too," continued Miss Baffin. "He proposed to me this morning."

"What! Proposed to you! Incredible! Why, the boy has not known you more than an hour or two."

"He is a man, pa; not a boy," said Miss Baffin, a little hurt. "It *was* rather sudden; but, then, genuine affection sometimes manifests itself in that way."

The Professor smiled; he perceived the exact situation of things. Then he looked very serious again. This was a contingency of which he had not taken account.

"Well, Tilly," he said, "I hardly know what to say about the matter. It is so completely unexpected. You didn't accept him?"

"No; not exactly, but—"

"Very well, then. We will leave the situation as it is for the present. When we have been here longer we can better determine what we should do."

Sir Dinadan entered with the Baron. The Baron greeted his guests with warmth, making no allusion to the occurrences in the Hermit's cell, and appearing, indeed, to have forgotten them.

"It is enough, sir, and fair damsel, that misfortune has thrown you upon our shores. You shall make this your home while you live."

"A thousand thanks," responded the Professor.

"I cherish the belief that I can be of service to you. By the way, may I ask how is the noble Lady Bors?"

"Suffering greatly. My son tells me you are a wise leech, and can give her release from her pain."

"I hope I can. If you will permit my daughter, here, to see the lady and to follow my directions, we may be able to help her."

"There," said the Baron, waving his hand, "are your apartments. When you have made ready we will summon you to our banquet."

"Your property, which was upon the beach, will be placed before you very soon," said Sir Dinadan.

The Professor and Miss Baffin entered the rooms, and the Baron withdrew with his son.

When the trunks came and were opened, the guests arrayed themselves in their finest costumes, and Miss Baffin contrived to give to her beauty a bewildering effect by an artistic arrangement of frippery, which received its consummation when she placed some lovely artificial flowers in her hair.

Then the Professor, giving her certain plasters and a soothing drug or two, requested a servant, who stood outside the door, to announce to Lady Bors that Miss Baffin was ready to give her treatment.

Sir Dinadan came forward and gallantly escorted Miss Baffin to his mother's room; where, after presenting her, he left her and returned to the Professor.

The young man led the Professor about the castle, showing him its apartments, its furniture and decorations, with an earnest purpose to try to find favour in the eyes of the father of the woman he loved. The Professor, for his part, was charmed with his companion, and his interest in the castle and its appurtenances increased every moment.

"This," said Sir Dinadan, pausing before a large oaken door, barred with iron, "is the portal to the upper room of the south tower. In this chamber the Baron has confined Ysolt, my sister, until she consents to think no more of Sir Bleoberis."

"Locked her up, has he? That seems hard."

"Cruel, is it not?"

"You favour the suit of the Knight, do you?" inquired the Professor.

"I would let Ysolt choose for herself. He is a worthy man; but he has poverty."

"We must try to help him," said the Professor.

"You would act differently in such a case; would you not?" asked Sir Dinadan, rather eagerly.

"Why, yes, of course; that is, I mean," said the Professor, suddenly recollecting himself, and what Miss Baffin had told him, "I mean, I would think about it. I would give the matter thoughtful consideration."

Sir Dinadan sighed, and asked the Professor if he would come with him to the dining-hall.

It was a noble room. As the Professor entered it with Sir Dinadan, as he looked at the vast fireplace filled with burning logs, because the air of the castle was chilly even in summer time, at the rudely carved beams that traversed the ceiling, at the quaint curtains and curious ornaments upon the walls, at the long table which stretched across the floor and bore upon its polished surface a multitude of vessels of strange and often fantastic shapes, he could hardly believe his senses. These things, this method of existence, he had read about myriads of times, but they had never seemed very real to him until he encountered them here face to face.

These people among whom he had come by such strange mischance actually lived and moved here, amid these scenes, and they were as common and as prosy to them as the scenes in his own home in the little enclosure hard by the walls of the university building at Wingohocking.

It was that home and its equipment that seemed strange and incongruous to him now. As he thought about it, he felt that he would experience an actual nervous shock if he should suddenly be plumped down in his own library. Very oddly, as his mind reverted to the subject, his memory recalled with peculiarly vivid distinctness an old and faded dressing-gown in which he used to come to breakfast; and a blue cream-jug with a broken handle, which used to be placed before him at the meal.

It seemed to him that the dressing-gown and the defective jug were as far back in the misty past as such a social condition as that with which he had now been brought into contact would have seemed if he had thought of it a month ago.

As the servants entered, bearing the viands upon large dishes, the Baron made his appearance at the upper end of the room, and a moment later Lady Bors walked slowly in, leaning upon the arm of Miss Baffin.

"Your sweet daughter," she said, when the Professor had been presented to her, "has eased my pain already. I think she must be an angel sent to me by Heaven."

"She *is* an angel," said Sir Dinadan, emphatically, so that his mother looked at him curiously. Miss Baffin blushed.

"Angels, my lady, do not come with porous plasters," said the Professor, smiling.

"I love her already, whether she is angel or woman," replied Lady Bors, patting Miss Baffin's arm.

"So do—," Sir Dinadan did not complete the sentence. It occurred to him that he might perhaps be getting a little too demonstrative.

"The Lady Tilly," said the Baroness, "has told me something of the adventure which brought you here. Will you be so courteous as to tell us more, and to inform us of that strange and wonderful land from which you have come?"

"Willingly, madam," replied the Professor. And so, while the meal was in progress, the Professor,—not neglecting the food, for he was really hungry,—tried, in the plainest language he could command, to convey to the minds of his hearers some notion of the marvels of modern civilization. The Baron, Lady Bors, and Sir Dinadan asked many questions, and they more than once expressed the greatest astonishment at the revelations made in the Professor's narrative.

"I will show you some of these wonders," said Professor Baffin. "Most happily I have with me in my trunks quite a number of instruments, such as those I have told you of."

"In your trunks!" exclaimed the Baron. "You do not wear trunks, as we do."

The Professor at once explained the misapprehension. When he had done, there was heard in the room the twanging of the strings of a rude musical instrument.

"It is the minstrel," said Sir Dinadan, as the Professor and Miss Baffin looked around.

The Professor was delighted.

"He is going to sing," said the Baron.

The bard, after a few preliminary thrums upon an imbecile harp, burst into song. He occupied several moments in reciting a ballad of chivalry, and although his manner was dramatic, his voice was sadly cracked and out of tune.

"Tilly," said the Professor, "remember to note in your journal that the musical system here is constructed from a defective minor scale, with incorrect intervals. I observed precisely the same characteristics in the song that our Irish nurse, Mary, used to put you to sleep with when you were a baby. I stood outside the chamber door one night, and wrote the strain down as she sang it. This proves that it is very ancient."

"You like the song, then?" asked the Baron.

"It is very interesting, indeed—very!" replied the Professor. "I think we shall obtain a great deal of valuable information here. No, Tilly, you had better refuse it," said the Professor, observing that Sir Dinadan, who appeared to be animated by a resolute purpose to stuff Miss Baffin, was pressing another dish upon her, "you will spoil your night's rest."

"Do you sing, Sir Baffin?" inquired Lady Bors.

"Never in company, my lady," replied the Professor; "my vocalization would excite too much alarm."

The Baron and his wife manifestly did not comprehend the pleasantry.

"My daughter sings very nicely; but you can hear her sing without her lips being opened. Excuse me for a moment."

The Professor went to his apartment, and presently returned, bringing with him a phonograph. Placing it upon the table, he turned the crank. From the funnel at once issued a lovely soprano voice, singing, with exquisite enunciation and inflection, a song, every word of which was heard by the listeners.

Lady Bors looked scared, Sir Dinadan crossed himself, the Baron eyed the Professor doubtfully, the minstrel over in the corner laid down his harp, and relieved his overcharged feelings by bursting into tears, which he wiped away with the sleeve of his tunic.

"It must be magic," said the Baron, at last; "no mere man could hide an angelic spirit in such a place, and compel it to sing."

"Allow me to explain," said the Professor; and then he unfolded the mechanism, and showed the method of its operation. "My daughter sang up several songs for me before we left home. They were stored away here for future use. Tilly, my love, sing something, so that our friends can perceive that it is the same voice."

Miss Baffin, after some hesitation, began "The Last Rose of Summer". While she sang, Sir Dinadan looked at her with rapture depicted on his countenance. When she had done he reflected for an instant, and then, rising and walking over to the place where the minstrel sat, he seized by the ear that unfortunate operator with defective minor scales, and, leading him to the door, he kicked him into the hall.

This appeared to relieve Sir Dinadan's feelings.

When he returned, the Professor persuaded him to have his voice recorded by the phonograph; and by the time the Baron and Lady Bors had also tried the experiment, the faith of the family in the powers of Professor Baffin had risen to such a pitch that the Baron would have been almost ready to lay wagers in favour of his omnipotence.

The Professor that evening accepted for himself and his daughter a very urgent invitation to make the castle their home, at least until Fate and the future should determine if they were to remain permanently upon the island. The chance that they would ever escape seemed indeed, exceedingly slender; and the Professor resolved to accept the promise with philosophical resignation.

He employed much of his time during the first weeks that he was the Baron's guest in making the Baron familiar with some of the wonders of modern discovery and invention. The Baron also was deeply interested in an exhibition given by the Professor of the powers of his patent india-rubber life-raft, which the Professor brought up from the beach folded into a small bundle. After inflating it, to the amazement of the spectators, he put it into the fosse that surrounded the castle and paddled about upon it. The raft was allowed to remain in the ditch ready for use.

The Professor often went outside the castle walls to talk with Sir Bleoberis, and to comfort him. The Professor explained the telegraph and the locomotive to the Knight; and when the Knight assured him that the armorers of the island could make the machinery that would be required, if they should receive suitable instructions, the Professor arranged to build a short railroad line and a telegraph line in partnership with Sir Bleoberis, if the latter would obtain the necessary concession from King Brandegore. Professor Baffin was of the opinion that the Knight, by such means, might ultimately acquire great wealth.

Meantime Sir Dagonet had been seen several times of late in the vicinity of the castle, and once he had made again a formal demand upon the Baron for Ysolt's hand. This the Baron refused, whereupon Sir Dagonet returned an insolent reply that he would have her in spite of her father's objection. The Professor sincerely pitied both Ysolt and Sir Bleoberis, but as the Baron always became violently angry when the suffering of the lovers was alluded to, the Professor disliked to plead their cause.

It occurred to him, however, one day that there could be no possible harm in arranging to permit the forlorn creatures to converse with each other; and so, with the help of Miss Baffin, who was allowed to enter the captive's room, he fixed up a telephone, the machinery of which he had in one of his trunks, with a wire running from Ysolt's window to a point some distance beyond the castle wall.

The battery with which the instruments were supplied was placed in an iron box furnished by Sir Bleoberis, and hidden behind a huge oak tree.

The lovers were delighted with the telephone and its performances; but the Professor's ingenious kindness caused him a great deal of serious trouble.

It seems that Miss Baffin one morning had been showing her father's umbrella to Ysolt, and making her acquainted with its peculiarities and uses.

When Miss Baffin had withdrawn, Sir Bleoberis began to breathe through the telephone protestations of his undying love, and finally he appealed to Ysolt to fly with him. Of course he expected nothing to come of this appeal, for he had not the slightest conception of any method by which Ysolt could escape from her prison. He merely threw it in, in a general sort of a way, as an expression of the intensity of his affection.

But it suggested to the mind of Ysolt an ingenious thought; and she responded through the telephone that if Sir Bleoberis would keep out of sight and have his gallant steed ready, she would join him in a few moments. The Knight's heart beat so fiercely at this news that it fairly made his armour vibrate.

Obeying the orders of Ysolt, he went behind the oak and sat upon the iron box containing the Professor's battery and electrical apparatus.

Ysolt's window was but twenty feet from the surface of the water in the fosse. Directly beneath it, by a most fortunate chance, floated the life-raft of

Professor Baffin. The brave girl, climbing upon the stone sill of the window, hoisted the umbrella, and sailing swiftly downward through the air, she alighted safely upon the raft. A single push upon the wall sent it to the further side of the ditch, whereupon Ysolt leaped ashore, unperceived by the warder or by any one in the castle.

A moment more, and seated upon the steed of her cavalier, with his strong arm around her, she would be flying to peace and happiness and love's sweet fulfilment, far, far beyond the reach of the angry Baron's power.

But, alas, human life is so full of mischances! As Ysolt neared the great oak behind which her lover sat, Sir Dagonet came riding carelessly across the lawn. Seeing her he spurred his horse forward, and, right before the eyes of Sir Bleoberis, he grasped her by the arm, tossed her to his saddle and dashed away across the country.

But why did not Sir Bleoberis leap to the rescue?

Sir Bleoberis tried with all his might to do so; but he had on a full suit of steel armour, and the Professor's battery, by some means even yet unexplained, so charged the cover of the box with magnetism that it held the Knight close down. He could not move a muscle of his legs. He writhed and twisted and expressed his fury in language that was vehement and scandalous; but the Professor's infamous machine held him fast; and he was compelled to sit by, imbecile and raging, while the wind bore to his ears the heart-rending screams of his sweetheart as she cried to him to come and save her from an awful fate.

The shrieks of the unhappy Ysolt penetrated to the castle, and at once the Baron ran out, followed by Sir Dinadan, Professor Baffin, and a host of the Baron's retainers, all of them armed and ready for war. The first act of the Professor was to capture his expanded umbrella, which was being blown about wildly by the wind. Furling it, he proceeded to the place where Sir Bleoberis sat, trying to explain to the infuriated Baron what had happened.

"There!" said Sir Bleoberis, savagely, pointing to the Professor, "is the vile wretch that did it all! Seize him! He, he alone is to blame."

The Professor was amazed.

"Yes!" exclaimed Sir Bleoberis, "it was he who persuaded the fair Ysolt to leap from the window; it was he who notified Sir Dagonet, and it is his wicked enchantment that held me here so that I could not fly to her succour. I cannot even get up now."

"The man," said the Professor to the Baron, "appears to be suffering from intellectual aberration. I can't imagine what he means. Why don't you rise?"

"You, foul wizard, know that I am held here by your infernal power!"

"Try to be calm," said the Professor, soothingly. "Your expressions are too strong. Let me see—. Why, bless my soul, the electrical current has magnetized the box. There, now," said the Professor as he snipped a couple of the wires, "try it again."

Sir Bleoberis arose without effort. Baron Bors stepped forward and said sternly:

"What, you, Sir Bleoberis, were doing here I do not know. I suspect you of evil purposes. But it is clear you had nothing to do with the seizure of my daughter, if, indeed, she has been carried off by Sir Dagonet. You may go. But as for you," shouted the Baron, turning to the Professor, "I perceive that your devilish arts have been used against me and my family while you have been eating my bread. The world shall no longer be burdened by such a monster. Away with him to the scaffold!"

"This," said the Professor, as the perspiration stood in beads upon his pallid face, "is painful; very painful. Allow me to explain. The fact is I—"

"Away!" said the Baron, with an impatient gesture. "Off with his head as quickly as possible!"

"But, my dear sir," contended the Professor, as the Baron's retainers seized him, "this is simply awful! No court, no jury, no trial, no chance to tell my story! It is not just. It is not fair play. Permit me, for one moment, to—"

"To the block with him!" screamed the Baron. "Have no more parley about it!"

Sir Bleoberis came forward.

"Sir Bors," he said, "this, in a measure, is my quarrel. It falls to me by right to punish this wretch. Will you permit me?" and then Sir Bleoberis struck the Professor in the face with his mailed gauntlet.

Professor Baffin would have assailed him upon the spot, but for the fact that he was a captive.

"He means that you shall fight him," said Sir Dinadan, who retained his faith in the Professor, remembering his own affection for Miss Baffin.

"Certainly I will," said the Professor. "Where, and when, and how? I would like to have it out right here on the spot."

It is melancholy to think what would have been the sorrow of the members of the Universal Peace Society, of which the Professor was the first vice-president, if they could have observed the eagerness with which that good man seemed to long for the fray, and the fiery rage which beamed from his eyes until the sparks almost appeared to fly from his spectacles.

Miss Baffin at this moment rushed upon the scene, and in wild affright flung her arms about her father.

"The contest shall be made," said the Baron, sternly. "Unhand him!"

The Professor hurriedly explained the matter to Matilda, who sobbed piteously.

"You shall have my armour, my horse, and my lance," said Sir Dinadan in a kindly voice to the Professor. "Go and get them," he continued, speaking to some of the servants.

"Thank you," said the Professor. "I am much obliged. You are a fine young man."

"But, pa," said Miss Baffin through her tears, "surely you are not going to fight?"

"Yes, my love."

"And you a member of the Peace Society, too."

"I can't help it, my child. You may omit to note this extraordinary occurrence in your journal. The Society may as well remain in ignorance of it. But I must conform to the customs of the place."

"How can you ever do anything upon a horse, with armour and a lance? It is dreadful!"

"No, my child, it may perhaps be regarded as fortunate. For many years I have longed to observe the practices of ancient chivalry more closely; that opportunity has now come. I am about to have actual practical experience with them."

Miss Baffin wiped her eyes as Sir Dinadan came to her side and tried to comfort her. Sir Agravaine, who had ridden up during the excitement, dismounted when he saw Miss Baffin, and pulling Sir Dinadan by the sleeve, he whispered:

"You are acquainted with that lady?"

"Yes."

"Would you mind ascertaining for me if I am to understand her remarkable conduct to me as tantamount to a refusal? I don't want to trouble you, but—"

Sir Dinadan turned abruptly away, leaving Sir Agravaine still involved in doubt.

When the armour came, Sir Dinadan helped the Professor to put it on. It was a size or two too large for him, and the Professor had a considerable amount of difficulty in adjusting the pieces properly, but, with the help of Sir Dinadan, he at last succeeded.

"Bring me my lance!" he exclaimed, with a firm voice, as he stepped forward.

"It is here," said Sir Dinadan.

"Farewell, my child," said the Professor to Miss Baffin, making a futile attempt to bend his elbows so that he could embrace her. "Farewell!" and the Professor tried to kiss her, but he merely succeeded in injuring her nose with the visor of his helmet.

"O pa!" said Miss Baffin, weeping, "if you should be killed."

"No danger of that love, none at all. I am perfectly safe. I feel exactly as if I were a cooking-stove, to be sure; but you may depend upon my giving a good account of myself. And now, dear, adieu! Ho, there!" exclaimed the Professor, with faint reminiscences of the tragic stage coming into his mind. "Bring me my steed!"

The determined efforts of four muscular men were required to mount the Professor upon his horse. And when he was fairly astride, with his lance in his hand, he felt as if he weighed at least three thousand pounds, and the weapon seemed quite as large as the jib-boom of the *Morning Star*.

The warrior did his best to sit his horse gracefully; but the miserable beast pranced and curveted in such a very unreasonable manner that his spectacles were continually shaking loose, and in his efforts to fix them, and at the same time to hold his horse, he lost control of his lance, and came near impaling two or three of the spectators.

Sir Dinadan's own groom then took the bridle-rein, and leading the horse quietly to the jousting-ground put him in place directly opposite to Sir Bleoberis, whose lance was in rest, and who evidently intended to spit the Professor through and through at the first encounter.

The Professor really felt uncomfortably at a disadvantage in his iron-clad condition, and he began to think that the sports and combats of the olden time were perhaps not so interesting after all, when brought within the range of practical experience.

Suddenly the herald's trumpet sounded a blast. The Professor had not the least notion of the meaning of the sound, but Sir Bleoberis started promptly towards him, and the Professor's horse, trained at jousting, also started. The Professor was not quite ready, and he pulled the rein hard while trying to fix his lance in its rest. This caused the horse to swerve sharply around, whereupon the warrior's spectacles came off, and the horse dashed at full speed to the side of the jousting-ground, bringing the half-blinded Professor's lance up against a tree, into which the point stuck fast. The Professor was hurled with some violence to the ground, and the horse ran away.

When they picked him up and unlatched his helmet, he was bleeding at the nose.

"It is of no consequence, Matilda, of no consequence, I assure you," he said. "I am shaken up a little, but not hurt. I think, perhaps, I need practice at this kind of thing."

The Professor, while speaking, felt about him in a bewildered way for the pocket in which he was used to keep his handkerchief. But as the armour baffled his efforts to find it, Miss Baffin offered him her kerchief with which to stanch the blood.

"The ancients, Matilda," said the Professor, as he pressed the handkerchief to his nose, "must have possessed great physical strength, and they could not have been near sighted. By the way, where are my glasses?"

Sir Dinadan handed them to him.

"You will not attempt to get on that horrid horse, again, pa, will you?" said Miss Baffin, entreatingly.

"I think not, my child, unless I am forced to do so. Jousting is interesting to read about; but as a matter of fact it is brutal. I think, Sir Dinadan, I should be more comfortable if I could get this cast-iron overcoat off, so that I could move my elbows without creaking."

Sir Dinadan helped him to remove his armour, and said:

"My noble mother has insisted that Sir Bleoberis shall not fight with you, and the Baron has yielded to her wish."

"How can I thank you?" exclaimed Miss Baffin.

Sir Dinadan looked at her as if he would like to tell her how, if he dared venture. But he only said:

"I deserve no thanks. My mother is upon your side and that of your father. She asks me to bring him to her."

The Baron was with his wife, and Sir Bleoberis stood before them.

"Sir Baffin," said the Baron, "Lady Bors insists that you are innocent of any wrong-doing; and Sir Bleoberis, seeing that you are unskilled, has resolved not to have a combat with you. I am willing to pardon you upon one condition: that you find my daughter and bring her back to me."

"That I should be willing to try to do under any circumstances," said the Professor. "I regret her loss very deeply. But, you see, I know nothing of the country. I am afraid I should not discover her if I should go alone."

"I will go with you," said Sir Bleoberis.

"That is first-rate," said the Professor. "Give me your hand."

"We will keep your daughter in the castle as a hostage," said the Baron. "When you return with Ysolt you shall have the Lady Tilly, and Sir Bleoberis shall have Ysolt."

"I am profoundly grateful," replied Sir Bleoberis, bowing.

"My dear," said the Professor to Miss Baffin, "does the arrangement suit you?"

"It suits me," muttered Sir Dinadan.

"I must stay whether I wish to or not," replied Miss Baffin. "But I shall worry about you every moment while you are gone."

"Sir Dinadan may be able to soothe her," said Sir Bleoberis, with a smile.

"I think I could, if I were allowed to try," insinuated Sir Agravaine.

"I charge Sir Dinadan and his noble parents with the task," said the Professor.

The entire party, with the exception of Sir Agravaine, then returned to the castle, so that the Professor could make ready for the journey.

Chapter III. The Rescue

Professor Baffin politely declined to wear the armour of Sir Dinadan upon the journey. He packed a few things in a satchel, and putting his revolver in his pocket, he bade adieu to his daughter and the members of the Baron's family. Mounting his horse by the side of Sir Bleoberis, who rode in full armour, the two trotted briskly out through the woods to the roadway, which ran by not far from the castle.

"Where shall we go to look for the lady?" asked the Professor, as the Knight started down the road at a rapid pace.

"The villain, no doubt, has carried her captive to his castle. We shall seek her there."

"How are we going to get her out? I have had very little experience, personally, in storming castles."

"We shall have to devise some plan when we get there," replied the Knight. "The castle, unhappily, is upon an island in the middle of the lake."

"And I can't swim," said the Professor.

"Perhaps the King will give us help. It is close to the place where he holds his court."

The Professor began to think that the case looked exceedingly unpromising. He lapsed into silence, thinking over the probable results of the failure of his mission; and as the Knight appeared to be absorbed in his own reflections, the pair rode forward without engaging in further conversation.

Professor Baffin did not fail to notice the extreme loveliness of the country through which they were passing. It presented all the characteristics of a perfect English landscape; but he observed that it was not fully cultivated, and that the agricultural methods employed were of a very primitive kind.

After an hour's ride, the two horsemen entered a wood. Hardly had they done so before they heard, near to them, the voice of a woman crying loudly for help. Sir Bleoberis at once spurred his horse forward, and the Professor followed close behind him.

Presently they perceived a Knight in armour endeavouring to hold upon the horse in front of him a young woman of handsome appearance, who screamed loudly as she attempted to release herself from his grasp.

"Drop her!" exclaimed the Professor in an excited manner, and drawing his revolver, "put her down; let her go at once!"

The Knight turned, and seeing the intruders he released the maiden, and levelling his lance, made straight for Sir Bleoberis at full gallop.

The lady, white with terror, flew to the Professor, and reposed her head upon his bosom.

Professor Baffin was embarrassed. He had no idea what he had better do or say. He could not repulse the poor creature; and as the situation, upon the whole, was not positively disagreeable, he permitted her to remain, sobbing upon his bosom, while he watched the fight and dried her eyes, in a fatherly way, with his handkerchief.

The two Knights came together with a terrible shock which made the sparks fly; but neither was unhorsed or injured, and the lances of both glanced aside. They turned, and made at each other again. This time the lance of each pierced the armour of the other, so that neither lance could be withdrawn. It really seemed as if the two knights would have to undress and to walk off, leaving their armour pinioned together. A moment later the strange Knight fell to the ground, and lay perfectly still. The Professor went up to him and taking his lance from his hand, so that Sir Bleoberis could move, unlaced the Knight's helmet.

He was dead.

The Professor was inexpressibly shocked. "Why," he exclaimed, "the man is dead! Most horrible, isn't it?"

"Oh, no," said Sir Bleoberis, coolly. "I tried to kill him."

"You wanted to murder him?"

"Oh, yes, of course."

"I am so glad you did," exclaimed the damsel with a sweet smile. "How can I thank you? And you, my dear preserver."

"Bless my soul, madam," exclaimed the Professor, "I had nothing to do with it. I consider it perfectly horrible."

Turning to Sir Bleoberis, the maiden said, "It was you who fought, but it was this brave and wise man who brought you here, was it not?"

"Yes," said Sir Bleoberis, smiling.

"I knew it," exclaimed the lady, flinging her arms around the Professor's neck. "I can never repay you—never, never, excepting with a life of devotion."

The Professor began to feel warm. Disengaging himself as speedily as possible, he said—

"Of course madam, I am very glad you have been rescued—very. But I deeply regret that the Knight over there was slain. What," asked the Professor of Sir Bleoberis, "will you do with him?"

"Let him lie. He is of no further use."

"I never heard of anything so shocking," said Professor Baffin. "And how are we to dispose of this lady?"

"I will go with you," exclaimed the damsel, looking eagerly at the Professor. "Let me tell you my story. My name is Bragwaine. I am the daughter of the Prince Sagramor. That dead Knight found me, a few hours ago, walking in the park by my father's castle. Sir Lamorak, he was called. Riding up swiftly to me, he seized me, and carried me away. He brought me, despite my screams and struggles, to this place, where you found us both. I should now be a captive in his castle but for you."

Bragwaine seemed about to fall upon the Professor's neck again, but he pretended to stumble, and retreated to a safe distance.

"Is there much of this kind of thing going on,—this business of galloping off with marriageable girls?" asked the Professor.

"Oh yes," said Sir Bleoberis.

"I thought so," said the Professor; "this is the second case I have encountered today. We shall most likely have quite a collection of rescued damsels on our hands by the time we get back home. It is interesting, but embarrassing."

"I know Prince Sagramor," said Sir Bleoberis to Bragwaine. "We are going to the court, and will take you to your father."

"*You* will take me, Sir—Sir—"

"Sir Baffin," explained Sir Bleoberis.

"Sir Baffin, will you not?"

"You can have my horse. I will walk."

"I will ride upon your horse with you, and you shall hold me on," said Bragwaine.

"That is the custom," said Bleoberis.

"But," exclaimed the Professor with an air of distress, "I am not used to riding double. I doubt if I can manage the horse and hold you on at the same time."

"You need not hold me," said Bragwaine laughingly; "I will hold fast to you. I shall not fall."

"But then—"

"I *will* go with you," said Bragwaine almost tearfully. "You won me from the hands of that villain, Lamorak, and I am not so ungrateful as to leave you to cling to another person."

"Well, I declare!" exclaimed the Professor, "this certainly is a very curious situation for a man like me to find himself in. However, I will do the best I can."

Professor Baffin mounted his steed, and then Sir Bleoberis swung the fair Bragwaine up to a place on the saddle in front of the Professor. Bragwaine clutched his coat-sleeve tightly; and although the Professor felt that there was no real necessity that she should attempt to preserve her equipoise by pressing his shoulder strongly with her head, he regarded the arrangement without very intense indignation.

He found that he could ride very comfortably with two in the saddle, but he felt that his attention could be given more effectively to the management of the horse if Bragwaine would stop turning her eyes up to his in that distracting manner so frequently.

They rode along in silence for awhile. Suddenly Bragwaine said:

"Sir Baffin?"

"Well; what?"

"Are you married?"

Professor Baffin hardly knew what answer he had better give. After hesitating for a moment, he said:

"I have been."

"Then your wife is dead?"

The Professor could not lie. He had to say "Yes!"

"I am so glad," murmured Bragwaine. "Not that she is dead, but that you are free."

Professor Baffin was afraid to ask why. He felt that matters were becoming serious.

"And the reason is," continued Bragwaine, "that I have learned to love you better than I love any other one on earth!"

She said this calmly, very modestly, and quite as if it were a matter of course.

The Professor in astonishment looked at Sir Bleoberis, who had heard Bragwaine's words. The Knight nodded to him pleasantly, and said, "I expected this."

Evidently it was not an unusual thing for ladies so to express their feelings.

The somewhat bewildered Sir Baffin then said, "Well, my dear child, it is very kind indeed for you to regard me in that manner. I have done nothing to deserve it."

"You are my rescuer, my benefactor, my heart's idol!"

"Persons at my time of life," said the Professor, blushing, "have to be extremely careful. I will be a father to you, of course! Oh, certainly, you may count upon me being a father to you, right along."

"I do not mean that I love you as a daughter. You must marry me; you dear Sir Baffin." Then she actually patted his cheek.

Professor Baffin could feel the cold perspiration trickling down his back.

"I think," he said to Sir Bleoberis, "that this is, everything considered, altogether *the* most stupendous combination of circumstances that ever came within the range of my observation. It is positively distressing."

"You will break my heart if you will not love me," said Bragwaine, as if she were going to cry.

"Well, well," replied the bewildered Professor, "we can consider the subject at some other time. Your father, you know, might have other views, and,—"

"The Prince, my father, will overwhelm you with gratitude for saving me. I know he will approve of our marriage. I will persuade him to have you knighted, and to secure for you some high place at court."

"That," said the Professor, "would probably make me acutely miserable for life."

Within an hour or two after the fight with Sir Lamorak, the Professor and his companions drew near to Callion, the town in which King Brandegore held his court.

Just before entering it they encountered Prince Sagramor coming out with a retinue of knights in pursuit of Sir Lamorak and his daughter. Naturally he was filled with joy at finding that she had been rescued and brought back to him.

After embracing her, he greeted Sir Bleoberis and the Professor warmly, thanking them for the service they had done to him. Bragwaine insisted upon the Professor's especial title to gratitude, and when she had told with eloquence of his wisdom and his valour, and had added to her story Sir Bleoberis's explanation of the Professor's adventures, the Prince saluted the latter, and said:

"There is only one way in which I can honour you, Sir Baffin. I perceive that already you have won the heart of this damsel. I had intended her for another. But she is fairly yours. Take her, gallant sir, and with her a loving father's blessing!"

Bragwaine wept for happiness.

"But, your highness, if I might be permitted to explain—" stammered the Professor.

"I know!" replied the Prince. "You will perhaps say you are poor. It is nothing. I will make you rich. It is enough for me that she loves you, and that you return it."

"I cannot sufficiently thank you for your kindness," said the Professor, "but really there is a—"

"If you are not noble, the King will cure that. He wants such brave men as you are in his service," said the Prince.

"I am a free-born American citizen, and the equal of any man on earth," said the Professor proudly, "but to tell you the honest truth, I—"

"You are not already married?" inquired the Prince, somewhat suspiciously.

"I have been married; my wife is dead, and—"

"Then, of course, you can marry Bragwaine. Sir Colgrevance," said the Prince to one of his attendants, "ride over and tell the abbot that Bragwaine will wish to be married tomorrow!"

"Tomorrow!" shrieked the Professor. "I really must protest; you are much too sudden. I have an important mission to fulfil, and I must attend to that first, and at once."

Sir Bleoberis explained to the Prince the nature of their errand, and told him the Professor's daughter was held as a hostage until he should bring Ysolt back to Baron Bors.

"We will delay the wedding, then," said the Prince. "And now, let us ride homeward."

If it had not been for the heart-rending manner in which everybody regarded him as the future husband of Bragwaine, and for the extreme tenderness of that lady's behaviour toward him, the Professor would have enjoyed hugely his sojourn at the court. King Brandegore regarded him from the first with high favour, and the sovereign's conduct of course sufficed to recommend the Professor to everybody else. The Professor found the King to be a man of rather large mind, and it was a continual source of pleasure to the learned man to unfold to the King, who listened with amazement and admiration, the wonders of modern invention, science, and discovery.

With what instruments the Professor's ingenuity could construct from the rude materials at hand; he showed a number of experiments, chiefly electrical, which so affected the King that he ordered the regular court magician to be executed as a perfectly hopeless humbug; but Professor Baffin's energetic protest saved the unhappy conjurer from so sad a fate.

An extemporized telegraph line, a few hundred yards in length, impressed the King more strongly than any other thing, and not only did he make to Sir Bleoberis and the Professor exclusive concessions of the right to build lines within his dominions, but he promised to organize, at an early day, a raid upon a neighbouring sovereign, for the purpose of obtaining plunder enough to give to the enterprise a handsome subsidy.

Sir Dagonet did not come to court during the Professor's stay. But there, in full view of the palace, a mile away in the lake, was his castle, and in that castle was the lovely Ysolt.

The Professor examined the building frequently through his field-glasses, which, by the way, the King regarded with unspeakable admiration; and more than once he thought he could distinguish Ysolt sitting by the window of one of the towers overlooking the lake.

The King several times sent to Sir Dagonet messages commanding Sir Dagonet to bring the damsel to him, but as Sir Dagonet invariably responded by trying to brain the messenger or to sink his boat, the King was forced to give it up as a hopeless case. Storming the castle was out of the question. None of the available boats were large enough to carry more than half a dozen men, and Sir Dagonet had many boats of great size which he could man, so as to assail any hostile fleet before it came beneath the castle wall.

But the Professor had a plan of his own, which he was working out in secret, while he waited. Sir Bleoberis had procured several skilful armorers, and under the directions of the Professor they undertook to construct, in rather a crude fashion, a small steam-engine. This, when the parts were completed, was fitted into a boat with a propeller screw, and when the craft was launched upon the lake, the Professor was delighted to find that it worked very nicely. The trial-trip was made at night, so that the secret of the existence of such a vessel might be kept from any of the friends of Sir Dagonet who might be loitering about.

It devolved upon Sir Bleoberis, by bribing a servant of Sir Dagonet's who came ashore, to send a message to Ysolt. She was ordered to watch at a given hour upon a certain night for a signal which should be given from a boat, beneath her window, and then to leap fearlessly into the water.

The night chosen was to be the eve of the Professor's wedding-day. The more Prince Sagramor saw of Professor Baffin and his feats, the more strongly did he admire him; and in order to make provision against any accident which should deprive his daughter of marriage with so remarkable a man, the Prince commanded the wedding-day to be fixed positively, despite the remonstrances which the Professor offered somewhat timidly, in view of the extreme delicacy of the matter.

Upon the night in question, the Professor, at the request of the King, who was very curious to have an opportunity to learn from practical experience the nature of the thing which the Professor called "a lecture", undertook to deliver in the dining-room of the palace the lecture upon Sociology, which he had prepared for his course in England.

The room was packed, and the interest and curiosity at first manifested were intense; but the Professor spoke for an hour and three-quarters, losing his place several times because of the wretched character of the lights, and when he had concluded, he was surprised to discover that his entire audience was sound asleep.

At first he felt rather annoyed, but in an instant he perceived that chance had arranged matters in an extremely favourable manner.

It was within precisely half an hour of the time when he was to be in the boat under the window of Ysolt.

Stepping softly from the platform, he went upon tiptoe from the room. Not a sleeper awoke. Hurrying from the palace to the shore, he found Sir Bleoberis sitting in the boat, and awaiting him with impatience.

The Professor entered the craft, and applying a lighted match to the wood beneath the boiler, he pushed the boat away from the shore, and waited until he could get steam enough to move with.

A few moments sufficed for this, and then, opening the throttle-valve gently, the tiny steamer sailed swiftly over the bosom of the lake, through the intense darkness, until the wall of the castle, dark and gloomy, loomed up directly ahead.

A light was faintly burning in Ysolt's chamber in the tower, and the casement was open.

As the prow of the boat lightly touched the stones of the wall and rested, Sir Bleoberis softly whistled.

"I have always been uncertain," said the Professor to himself, "if the ancients knew how to whistle. This seems to indicate that they did know how. It is extremely interesting. I must remember to tell Tilly to note it in her journal."

In response to the signal, a head appeared at the casement, and a soft, sweet voice said:

"Is that you, darling?"

"Yes, yes, it is I," replied Sir Bleoberis. "Oh, my love! my Ysolt!" he exclaimed, in an ecstasy.

"Is Sir Baffin there, too?"

"Yes. We are both here; and we have a swift boat. Come to me at once, dear love, that we may fly with you homeward."

"I am not quite ready, love," replied Ysolt. "Will not you wait for a moment?"

"It is important," said the Professor, "that we should act quickly."

"But I *must* fix up my hair," returned Ysolt. "I will hurry as much as I can."

"Women," said the Professor to his companion, "are all alike. She would rather remain in prison for life than come out with her hair mussed."

The occupants of the boat waited very impatiently for fifteen or twenty minutes. Then Ysolt, coming again to the window, said:

"Are you there, dearest?"

"Yes," replied Sir Bleoberis, eagerly. "We are all ready."

"And there's no time to lose," added Professor Baffin.

"Is your hair fixed?" asked the Knight.

"Oh, yes," said Ysolt.

"Then come right down."

"Would ten minutes more make any difference?" asked Ysolt.

"It might ruin us," replied the Professor.

"We can wait no longer, darling," said Sir Bleoberis, firmly.

"Then you will have to go without me," said Ysolt, with a tinge of bitterness. "It is simply impossible for me to come till I get my bundle packed."

"We will wait, then," returned Sir Bleoberis, gloomily. Then he said to the Professor: "She had no bundle with her when she was captured."

The Professor, in silent desperation, banked his fires, threw open the furnace-door, and began to wonder what kind of chance he would have in the event of a boiler explosion. Blowing off steam, under the existing circumstances, was simply out of the question.

After a delay of considerable duration, Ysolt's voice was heard again:

"Dearest!"

"What, love?" asked Sir Bleoberis.

"I am all ready now," said Ysolt.

"So are we."

"How must I get down?"

"Climb through the window and jump. You will fall into the water, but I shall catch you and place you in the boat."

"But I shall get horridly wet!"

"Of course; but, darling, that can make no great difference, so that you escape."

"And spoil my clothes, too!"

"Yes, Ysolt, I know; but—"

"I cannot do it; I am afraid." And Ysolt began to cry.

Wild despair filled the heart of Sir Bleoberis.

"I have a rope here," said the Professor; "but how are we to get it up to her?"

"Ysolt," said Bleoberis, "if I throw you the end of a rope, do you think you can catch it?"

"I will try."

Sir Bleoberis threw it. He threw it again. He threw it thirteen times, and then Ysolt contrived to catch it.

"What shall I do with it now?" she asked.

"Tie it fast to something; to the bed, or anything," replied the Knight.

"Now what shall I do?" asked the maiden, when she had made the rope secure.

"Slide right down into the boat," said the Professor.

"It would ruin my hands," said Ysolt, mournfully.

"Make the attempt, and hold on tightly," said Sir Bleoberis.

"We shall be caught if we stay here much longer," observed the Professor, with anxious thoughts of the boiler.

"Goodbye then! I am lost. Go without me! Save yourselves! Oh, this is terrible!" Ysolt began again to cry.

"I will help her," said Sir Bleoberis, seizing the rope and clambering up the wall until he reached the window.

Day began to dawn as he disappeared in the room. The Professor started his fire afresh and shut the furnace-door. Sir Bleoberis, he knew, would bring down Ysolt without delay.

A moment later, the Knight seated himself upon the stone sill of the window and caught the rope with his feet and one of his hands. Then he placed his arm about Ysolt, lifted her out and began to descend.

Professor Baffin, even in his condition of intense anxiety, could not fail to admire the splendid physical strength of the Knight. When the pair were about half-way down, the rope broke, and Ysolt and Sir Bleoberis were plunged into the lake.

The Professor, excited as he was by the accident, remembered the boiler, and determined that he would have to blow off steam and take the consequences; so he threw open the valve, and instantly the castle walls sent the fierce sound out over the waters.

Sir Bleoberis, with Ysolt upon his arm, managed to swim to the side of the boat, and the Professor after a severe effort lifted her in. Then he gave his hand to the Knight, and as Sir Bleoberis's foot touched the side the Professor shut off steam, opened his throttle-valve, backed the boat away from the wall, and started for the shore.

It was now daylight. As the boat turned the corner of the wall, it almost came into collision with a boat in which, with ten oarsmen, sat Sir Dagonet. The inmates of the castle had been alarmed by the performances of the Professor's escape-pipe; and Sir Dagonet had come out to ascertain the cause of the extraordinary noise.

The Professor's presence of mind was perfect. Turning his boat quickly to the right, he gave the engine a full head of steam and shot away before Sir Dagonet's boat could stop its headway.

Sir Dagonet had perceived Ysolt, and recognized Sir Bleoberis. White with rage he screamed to them to stop, and he hurled at them terrible threats of vengeance if he should overtake them. As no heed was given to him he urged his rowers to put forth their mightiest efforts, and soon his boat was in hot pursuit of that in which the maiden, the Knight, and the Professor fled away from him.

By some means the people of the town of Callion had had their attention drawn to the proceedings at the castle, and now the shore was lined with spectators who watched with eager interest the race between Sir Dagonet's boat and the wonderful craft which had neither oars nor sails, and which sent a long streamer of smoke from out its chimney.

Professor Baffin, positively determined not to wed the daughter of Prince Sagramor, had prepared a stratagem. He had sent three horses to the side of the lake opposite to the town, and three or four miles distant from it, with the intention of landing there, and hurrying with Ysolt and Sir Bleoberis to the home of Baron Bors, without the knowledge of the Prince.

The daylight interfered, to some extent, with the promise of the plan, but Professor Baffin resolved to carry it out at any rate, taking what he considered to be the tolerably good chances of success. He turned the prow of his boat directly toward the town, making as if he would go thither. The pursuers followed fast, and as the Professor perceived that he could easily outstrip them, he slowed his engine somewhat, permitting Sir Dagonet to gain upon him.

When he was within a few hundred yards of the shore, close enough indeed, for him to perceive that the King, Prince Sagramor, Bragwaine, and all the attendants of the court were among those who watched the race with excited interest, the Professor suddenly turned his boat half around, and putting the engine at its highest speed, ploughed swiftly toward the opposite shore.

A mighty shout went up from the onlookers. Manifestly the fugitives had the sympathy of the crowd.

The oarsmen of Sir Dagonet worked right valiantly to win the chase, but the steamer gained constantly upon them; and when her keel grated upon the sand, close by where the horses stood, the pursuers were at least a third of a mile behind.

Sir Bleoberis sprang from the boat, and helped Ysolt to alight. The Professor stopped to make the fire in the furnace more brisk, and to tie down the safety valve; then hurrying after Sir Bleoberis and Ysolt, the three mounted their horses and galloped away.

In a few moments they reached the top of a hill which commanded a view of the lake. They stopped and looked back. Sir Dagonet had just touched the shore, but, as he had no horse, further pursuit was useless. So, shaking his fist at the distant party, he turned away with an affectation of contempt, and entered the Professor's boat to satisfy his curiosity respecting it.

"Let him be careful how he meddles with that," said the Professor.

As he spoke, the boat was torn to fragments. Sir Dagonet and two of his men were seen to fall, and a second afterwards the dull, heavy detonation of an explosion reached the ears of the Professor and his friends.

"It is dreadful," said the Professor with a sigh, "but self-preservation is the first law of nature, and then he had no right to run away with Ysolt, at any rate."

Chapter IV. How the Professor Went Home

The three friends turned their horses' heads away from the lake, and pressed swiftly along the road.

"It is necessary," said Professor Baffin, "that we should make good speed, for Prince Sagramor saw us come to this side of the lake, and if he shall suspect our design no doubt he will at once pursue us, in behalf of that abominable girl, his daughter."

The journey was made in silence during most of the time, for the hard riding rendered conversation exceedingly difficult, but whenever the party reached the crest of a hill which commanded a view of the road in the rear, the Professor looked anxiously behind him to ascertain if anybody was giving chase. When within a mile or two of Lonazep, he did at last perceive what appeared to be a group of horsemen at some distance behind him, and although he felt by no means certain that the Prince was among them, he nervously urged his companions forward, spurring, meantime, his own horse furiously, in the hope that he might reach the castle of Baron Bors ere he should be overtaken.

As the party came within sight of the castle, they could hear the hoofs of the horses of the pursuers, and soon their ears were assailed by cries, demanding that they should stop. It was, indeed, Prince Sagramor and his knights, who were following fast. The Professor galloped more furiously than ever when he ascertained the truth, and Sir Bleoberis and Ysolt kept pace with him.

Just as they reached the drawbridge, however, they were overtaken; and, as it was raised, they were compelled to stop and meet the Prince face to face. The Professor hurriedly called to the warder to lower the bridge, so that Ysolt could take refuge in the castle. Then he turned, and determined to make the best of the situation. The Prince was disposed to be conciliatory.

"We came," he said, "to escort you back again. We have a guard of honour here fitting for any bridegroom."

"You are uncommonly kind," replied the Professor, "but the parade is rather unnecessary. I am not going back just at present."

"I promised Bragwaine that you would return with us," said the Prince, sternly.

"Well, you ought not to make rash promises," replied the Professor, with firmness.

"You will go, of course?"

"Of course I will not go."

"Bragwaine is waiting for you."

"That," said the Professor, "is a matter of perfect indifference to me."

"I will not be trifled with, sir," said the Prince, angrily.

"Nor will I," exclaimed the Professor. "Let us understand one another. I do not wish to marry any one. I did not ask your daughter to marry me, and I have never consented to the union. I tell you now that I positively and absolutely refuse to be forced to marry her or any other woman. I will do as I please about it; not as you please."

"Seize him," shrieked the Prince to his attendants.

"Stand off," said the Professor, presenting his revolver. "I'll kill the man who approaches me. I shall put up with this foolishness no longer."

One of the knights rode toward him. The Professor fired, and the cavalier's horse rolled in the dust. The Prince and his people were stupefied with astonishment.

At this juncture, Baron Bors, Sir Dinadan, Sir Agravaine, Sir Bleoberis, and Miss Baffin emerged from the castle. Miss Baffin flew to her father, and flung her arms about him. The Professor kissed her tenderly, and as he did so, his eye caught sight of the wire of the telephone which he had arranged for Ysolt and Sir Bleoberis. A happy thought struck him. Advancing, he said to the Prince:

"It is useless for us to quarrel over this matter. Baron Bors has here an oracle. Let us consult that."

Then the Professor whispered something to Miss Baffin, who withdrew unobserved and went into the castle.

The Prince was at first indisposed to condescend to accept the offer, but his curiosity finally overcame his pride.

"Step this way," said the Professor. "Ask your questions through this," handing him the mouthpiece, "and put this to your ear for the answer."

"What shall I say?" inquired the Prince.

"Ask if it is right that I should marry your daughter."

The Prince put the question, and the answer came.

"What does the oracle say?" asked the Professor.

"It says you shall not," replied the Prince, looking a good deal scared.

"Are you satisfied?" said the Professor.

The Prince did not answer, but he looked as if he suspected a trick of some kind, and would like to impale Professor Baffin with his lance, if he dared.

He was about to turn away in disgust, when Sir Agravaine, who stood beside him, in a few half-whispered words explained to him the method by which the Professor had imposed upon him.

In a raging fury, the Prince rode up to the Professor, and would have assailed him; but Baron Bors advanced and said:

"This gentleman is unarmed, and unused to our methods of combat. He is my guest, and he has saved my daughter. I will fight his battles."

The Prince threw his glove at the Baron's feet. Baron Bors called for his armour and his horse, and when he was ready he took his place opposite to his antagonist, and waited the signal for the contest.

"This," said the Professor, "is probably the most asinine proceeding upon record. Because I won't marry Sagramor's daughter, Sagramor is going to fight with a man who never saw his daughter."

The combat was not a long one. At the first shock both knights were unhorsed; but, drawing their swords, they rushed together and hacked at each other until the sparks flew in showers from their armor.

The Baron fought well, but presently the Prince's sword struck his shoulder with a blow which carried the blade down through the steel plate, and caused the blood to spurt forth. The Baron fell to the earth; and Prince Sagramor, remembering the small number of his attendants, and the probability that he might be assailed by the Baron's people, mounted his horse and slowly trotted away without deigning to look at Professor Baffin. They carried the Baron tenderly into the castle, and put him to bed. The wound was a terrible one, and the Professor perceived that the chances of his recovery, under the rude medical treatment that could be obtained, were not very favourable. After doing what he could to help the sufferer, he withdrew from the room, and left the Baron with Lady Bors and the medical practitioner who was ordinarily employed by the family.

Miss Baffin, with Sir Dinadan, awaited her father in the hall. This was the first opportunity he had had to greet her. After some preliminary conversation, and after the Professor had expressed to Sir Dinadan his regret that the Baron should have been injured, the Professor said:

"And now, Tilly, my love, how have you been employing yourself during my absence?"

Miss Baffin blushed.

"Have you kept the journal regularly?" asked the Professor.

"Not so *very* regularly," replied Miss Baffin.

"I have a number of interesting and extraordinary things for you to record," said the Professor. "Has nothing of a remarkable character happened here during my absence?"

"Oh, yes," said Miss Baffin.

"I have learned to smoke," said Sir Dinadan.

"Indeed," said the Professor with a slight pang. "And how many cigars have you smoked?"

"Only one," replied the Knight. "It made me ill for two days. I think, perhaps, I shall give up smoking."

"I would advise you to. It is a bad habit," said the Professor, "and expensive. And then, you know, cigars are so dreadfully scarce, too."

"The Lady Tilly was very kind to me while I was ill. I believe I was delirious once or twice; and I was so touched by her sweet patience that I again proposed to her."

"While you were delirious?" asked the Professor.

"Oh, no; when I had recovered."

"What did you say to that, Tilly?" asked Professor Baffin.

"I referred him to you," replied Miss Baffin.

"But what will the Baron say?" asked the Professor.

"He and my mother have given their consent," said Sir Dinadan. "They declared that I could not have pleased them better than by making such a choice."

"Well, I don't know," said the Professor, reflectively. "I like you first-rate, and if I felt certain we were going to stay here—"

"I will go with you if you leave the island," said Sir Dinadan, eagerly.

"And then you know, Din," continued the Professor familiarly, "Tilly is highly educated, while you—Well, you know you must learn to read, and write, and cipher, the very first thing."

"I have been giving him lessons while you were away," said Miss Baffin.

"How does he get along?"

"Quite well. He can do short division with a little help, and he has learned as far as the eighth line in the multiplication table."

"Eight eights are sixty-four, eight nines are seventy-two, eight tens are eighty," said Sir Dinadan, triumphantly.

"Well," said the Professor, "if Tilly loves you, and you love Tilly, I shall make no objection."

"Oh, thank you," exclaimed both of the lovers.

"But, I tell you what, Din, you are getting a good bargain. There is no finer girl, or a smarter one either, on the globe. You people here cannot half appreciate her."

For more than a week, Baron Bors failed to show any signs of improvement, and the Professor thought he perceived clearly that his case was fast getting beyond hope. He deemed it prudent, however, to keep his opinion from the members of the Baron's family. But the Baron himself soon reached the same conclusion, and one day Lady Bors came out of his room to summon Sir Dinadan, Ysolt, Sir Bleoberis, who was now formally betrothed to Ysolt, and the Professor, to the Baron's bedside.

The Baron said to them, in a feeble voice, that he felt his end approaching, and that he desired to give some instructions, and to say farewell to his family. Then he addressed himself first to Sir Dinadan, and next to Ysolt. When he had finished speaking to them he said to Lady Bors,—

"And now, Ettard, a final word to you. I am going away, and you will need another friend, protector, companion, husband. Have you ever thought of any one whom you should like, other than me?"

"Never, never, never," said Lady Bors, sobbing.

"Let me advise you, then. Who would be more likely to fill my place in your heart acceptably than our good and wise and wonderful friend Sir Baffin?"

"Good gracious!" exclaimed the Professor with a start.

"Your son is to marry his daughter; and she will be happy to be here with him in the castle. Promise me that you will try to love him."

"Yes, I will try," said Lady Bors, wiping her eyes and seeming, upon the whole, rather more cheerful.

"That," said the Baron, "does not altogether satisfy me. I place upon you my command that you shall marry him. Will you consent to obey?"

"I will consent to anything, so that your last hour may be happier," said Lady Bors with an air of resignation. She was supported during the trial, perhaps, by the reflection that in dealing with lumbago Professor Baffin had no superior in the kingdom.

Father Anselm was announced. "Withdraw, now," said the Baron to all of his family but Lady Bors. "I must speak with the Hermit."

Professor Baffin encountered the Hermit at the door. The holy man stopped long enough to say that a huge ship had come near to the shore upon which the Professor had landed, and that it was anchored there. From its mast, Father Anselm said, fluttered a banner of red and white stripes with a starry field of blue.

The Professor's heart beat fast. For a moment he could hardly control his emotion. He resolved to go at once to the shore and to take his daughter with him. Withdrawing her from her companions the two strolled slowly out from the castle into the park. Then, hastening their steps, they passed towards the shore. In a few moments they reached it, and there, sure enough, they saw a barque at anchor, while from her mast-head floated the American flag.

A boat belonging to the barque had come to the shore to obtain water from the stream. Professor Baffin entered into conversation with the officer who commanded the boat. The vessel proved to be the *Mary L. Simpson*, of Martha's Vineyard, bound from the Azores to New York. When the Professor had explained to the officer that he and his daughter were Americans, the mate invited them to come aboard so that he could introduce them to the captain.

"Shall we go, my child?" asked the Professor.

"If we can return in a very few moments, we might go," said Miss Baffin.

They entered the boat, and when they reached the vessel, they were warmly greeted by Captain Magruder.

While they were talking with him in his cabin the air suddenly darkened, and the captain rushed out upon deck. Almost before he reached it a terrific gale struck the barque, and she began to drag her anchors. Fortunately the wind blew off shore, and the captain, weighing anchor, let the barque drive right out to sea. The Professor was about to remark to Miss Baffin that he feared there was small chance of his ever seeing the island again, when a lurch of the vessel threw him over. His head struck the sharp corner of the captain's chest, and he became unconscious.

When Professor Baffin regained his senses, he found that he was lying in a berth in a ship's cabin. Some one was sitting beside him,—

"Is that you, Tilly?" he asked, in a faint voice.

"Yes, pa; I am glad you are conscious again. Can I give you anything?"

"Have I been long unconscious, Tilly?"

"You have been very ill for several days; delirious sometimes."

"Is the captain going back to the island?"

"Going back to the *what*, pa?"

"To the island. It must have seemed dreadfully heartless for us to leave the castle while the Baron was dying."

"While the Baron was dying! What do you mean?"

"Why, Baron Bors could not have lived much longer. I am afraid Sir Dinadan will think hard of us."

"I haven't the least idea what you are talking about. Poor pa! your mind is beginning to wander again. Turn over, and try to go to sleep."

Professor Baffin was silent for a moment. Then he said,—

"Tilly, do you mean to say you never heard of Baron Bors?"

"Never."

"And that you were never engaged to Sir Dinadan?"

"Pa, how absurd! Who are these people?"

"Were you not upon the island with me, at the castle?"

"How could we have gone upon an island, pa, when we were taken from the raft by the ship?"

"Tilly, my child, when I get perfectly well I shall have to tell you of the most extraordinary series of circumstances that has come under my observation during the whole course of my existence!"

Then Professor Baffin closed his eyes and fell into a doze, and Miss Baffin went up to tell the surgeon of the ship *Undine*, from Philadelphia to Glasgow, that her father seemed to be getting better.

Max Adeler (1841–1915) was the pen name of the American novelist Charles Heber Clark. Adeler/Clark was primarily a humourist, becoming popular for his 1874 novel "Out of the Hurly Burly", but nevertheless wrote several stories of interest to science fiction readers. He is perhaps now best known for the events following publication of this 1880 novelette "The Fortunate Island": Clark claimed Mark Twain had plagiarized his story in Twain's 1889 novel "A Connecticut Yankee in King Arthur's Court".

Commentary

The literary subgenre in which travellers report fantastic encounters has a long history. In the second century AD, for example, Lucian of Samosata wrote *A True Story*, a work that ridiculed earlier travel writers who boasted of extraordinary adventures. For Lucian to believe this a worthwhile target for his satire, the subgenre must date back long before him. More than 1500 years later, Jonathan Swift published his (to my mind science-fictional) novel *Gulliver's Travels*. In his satire Swift of course took aim at heavyweight topics—politics, religion, human nature. But he also poked fun at less important targets, including that ancient trope of travellers chancing upon wondrous things. Most readers have perhaps always understood that such stories contain embellishments, if not outright lies. Less sophisticated readers, though, could accept the tales at face value. After all, when Lucian wrote *A True Story* much of Earth's surface remained unexplored. When Swift wrote *Gulliver's Travels*, most Europeans had at best only a vague idea of Earth's size. Why shouldn't explorers find lost worlds and lost tribes over the horizon?

By the time Adeler wrote "The fortunate island", people had a better understanding of the contours of Earth's continents. Authors had fewer plausible locations in which to situate a lost world. If they wanted to invoke the idea of Atlantis, then the action had better take place in a reverie or a fever dream. The motivation for writing 'lost-world' stories could have dried up. For a while, though, the popularity of these stories seemed to increase in inverse proportion to the decreasing area of unmapped territory. (A subset of 'lost-world' fiction, involving the crackpot idea of a hollow Earth, somehow endured for decades. Even as late as 1956, producers could make a movie such as *The Mole People*, the plot of which revolves around mountaineers falling through a hole in the Himalayas, entering a subterranean world, and meeting descendants of the Sumerian civilisation.)

Cartographers have now mapped every square kilometer of our planet's surface. We know that no holes lead to a hollow Earth. The oceans hold no islands like Lilliput, home to strange civilisations. The world contains no lost Atlantises for us to stumble upon. Given all this knowledge, one can sometimes forget that we still share our planet with uncontacted peoples. Some indigenous peoples have chosen to remain isolated from their neighbours and the wider world. Historical factors, cultural beliefs, or a desire to preserve a traditional way of life might all influence that choice. Most of these peoples live in the dense forests of South America, Central Africa, Papua New Guinea, and parts of southeast Asia. We know about them through occasional encounters with bordering tribes and drone footage. Survival International, an organisation that champions indigenous and tribal peoples' rights, suggests that more than 100 uncontacted tribes might still survive with a total population as high 10,000 individuals. (Fig. 4.1 shows members of an uncontacted tribe living in the state of Acre in Brazil.)

Fig. 4.1 Members of an isolated tribe in Acre, Brazil look up at a drone. Two men appear to aiming arrows at the drone. More uncontacted tribes live in the rainforests of Brazil than any other place on Earth. (Credit: Gleilson Miranda/Governo do Acre, CC BY 2.0 DEED Generic)

The fictional Professor Baffin would have had no scruples when dealing with uncontacted peoples (as would real-world explorers of the time). A casual, unexamined racism permeated those Victorian and Edwardian 'lost-world' stories. Today, we wrestle more with the ethical issues involved when making contact with people who want to remain apart. Such contact raises hard questions about respect for autonomy and prevention of harm. On the one hand, these groups have explicitly rejected outside contact. They seek to maintain their way of life. Forcing interaction would violate their right to self-determination. It would expose them to diseases for which they lack immunity. And it would risk the destruction of languages and cultures. On the other hand, complete non-interference comes with costs. Contact would provide these peoples with access to modern medicine. It would give them better protection against natural disasters. And it would supply them with technology to improve their quality of life. Should we make contact, then, or not? Reasonable people disagree about the right approach to take.

Several governments, mindful of the history of abuse suffered by smaller tribes at the hands of majority groups, have bestowed legal protection upon uncontacted tribes and their lands. But that protection sometimes fails to prevent illegal logging, mining, and drug trafficking in protected territories. And it doesn't stop individuals from trying to make contact. These interactions often do not end well, for one party or the other. Consider, for example, the case of the American missionary John Allen Chau and the Sentinelese. The Sentinelese people live on North Sentinel Island in the Bay of Bengal. They number perhaps 200 individuals and refuse to interact with the modern world. The government of India has adopted an 'eyes only' approach to the tribe. In 2018, Chau made an illegal visit to North Sentinel Island. He wanted to introduce the people there to Christianity. They killed him.

In Chap. 3 we discussed SETI and METI. The former program searches for extraterrestrial intelligence. The latter attempts to send a message to them. If we made first contact with ET, we would likely find their technology more advanced—probably *much* more advanced—than ours. How would we react to this mismatch in technological know-how? How would they react? Would we beg to know the secret of their stardrive? Or, like the Sentinelese people, would we try to remain aloof? Would we have any say in the matter? Those aliens *might* have their own version of *Star Trek*'s Prime Directive. They *might* avoid interference with other cultures and ecosystems, in the same way nation states here forbid interaction with unconnected peoples. As I recall, though, the crew of the *USS Enterprise* breached the Prime Directive more often than they observed it. Perhaps those aliens will make contact regardless of our feelings in the matter.

Of course, there might be no advanced aliens out there for the SETI program to detect. The METI project might be shouting into the void. Even in this case, though, might we still have to navigate the ethical questions around first contact (albeit in a less fraught context)?

Today, humans explore Mars—a planet that might once have been home to microbial life—using robotic rovers. In years to come, we will send astronauts. At some point, we will explore Europa (a moon of Jupiter) and Enceladus (a moon of Saturn) both of which have water and thus might host microbes. Perhaps life never started on Mars or Europa or Enceladus. But perhaps it did. Finding even primitive forms of extraterrestrial life on any of these worlds would revolutionize our understanding of life here on Earth. Our missions to these places, however, run the risk of contaminating pristine environments. Engineers construct robotic probes under sterile conditions, but the possibility of contamination remains. When astronauts start to explore these places so, inevitably, will microbes. Contamination from Earth life is almost certain. In that case, how do we balance the value of discovery with the imperative to do no harm?

Whether talking about uncontacted tribes or unexplored planets, we encounter the same dilemmas. They occur whenever a technologically advanced society encounters life that lacks technology. How should we interact in a respectful, minimally invasive way? Does scientific curiosity justify intervention in systems that have operated autonomously for centuries? (If ecosystems exist on Europa or Enceladus, then we must measure their isolation in aeons rather than centuries.) Should we feel obliged to restrain our influence when encountering the unconnected? Or, as some believe, do we have dominion over all living things so we can do whatever we wish?

Professor Baffin's encounter with the people of Atlantis turned out to be a fever dream. Neither he nor his author, Adeler, bothered to think of the ethics involved. Today, we grapple with the difficult questions surrounding interactions with uncontacted tribes. Tomorrow, as we explore more of the universe, those questions will persist.

Notes and Further Reading

Swift of course took aim at heavyweight topics—For an annotation and interpretation of Swift's masterwork, see Asimov (1980).
the popularity of these stories seemed to increase—Salmonson (n.d.) is a website with details on over 1500 titles on a theme related to lost worlds. Some of them most famous works include *King Solomon's Mines* (1885, by Rider Haggard); *The Man Who Would Be King* (1888, by Kipling); *The Lost World*

(1912, by Doyle); and *The Land That Time Forgot* (1918, by Burroughs). The 1956 movie *The Mole People* was directed by Virgil Vogel.

more than 100 uncontacted tribes—It is difficult to know how many such tribes exist. Even the definition 'uncontacted' is imprecise. It is almost impossible for tribes to avoid *all* contact with outsiders. People trade. When anthropologists do make contact with 'uncontacted' tribes they sometimes find them using utensils made in Chinese factories. I would not be surprised if the occasional Manchester United football shirt found its way to these peoples. Nevertheless, some groups try hard to avoid contact with others; some groups violently reject contact with others. Geographic isolation makes this possible. Survival International (n.d.) has more details.

ethical issues involved when making contact—For a discussion of the ethical principles involved with contacting uncontacted peoples, see Wong (2019).

humans explore Mars—See Uri (2022) for the story of 25 years of continuous robotic Mars exploration. At the time of this writing, the *Curiosity* and *Perseverance* rovers are operating on the red planet. Mann (2018) gives a brief introduction to the hunt for microbial life in the solar system.

References

Asimov, I.: The Annotated Gulliver's Travels. Potter, New York (1980)

Mann, A.: Inner workings: hunting for microbial life throughout the solar system. Proc. Natl. Acad. Sci. USA. **115**(45), 11348–11350 (2018). https://doi.org/10.1073/pnas.1816535115

Salmonson, J.A.: The Lost Race Checklist. www.rohpress.com/lost_race_check_guide.html (n.d.)

Survival International: Homepage. www.survivalinternational.org (n.d.)

Uri, J.: 25 years of continuous robotic Mars exploration—from Pathfinder to Perseverance. www.nasa.gov/history/25-years-of-continuous-robotic-mars-exploration-from-pathfinder-to-perseverance/ (2022)

Wong, B. (2019) Should we contact uncontacted peoples? A case for a Samaritan Rescue Principle. Oxford Uehiro Prize in Practical Ethics. Joint runner up. University of Oxford

5

Applications of Machine Learning

The Supersensitive Golf Ball (Edwin L. Sabin)

Beethoven Watkins, ordinarily as mild a mannered individual as ever endured with equanimity a crowded streetcar, swore dreadfully on the golf links.

Off the links, Watkins never had been known to use improper language. He was absolutely a gentleman—suave in demeanour, precise in deportment, and blameless in speech. That on the links he should be able to extemporize so fluently may be adduced as a proof of the tremendous mental strain that the individual pursuit of golf imposes upon its worshippers.

The passion that causes golfiacs to leave home, friends, and business in order to follow the ball is based upon two human characteristics—ambition and combativeness. The ambition may be social or athletic. The combativeness has only one objective point—that of subduing the little sphere, which, as all golfists know, is at times a living, breathing, devilish nucleus of malignity.

Now many a man forgets himself while playing golf, but as a rule golf swears are not considered genuinely abhorrent parts of speech. Rather are they condoned, if used in moderation—of course always in moderation. But the language of Watkins when he made a particularly bad stroke, or had a particularly bad lie, was so hearty, so blood-curdling, and bore such evidence of sincerity that hearers blushed. Watkins did not blush, because doubtless more than half the time he did not comprehend the import of his words.

After he had been in the golf club long enough for his peculiarity to become generally known, the other people avoided, if possible, playing with him. Everybody liked Watkins, and realized he was not vicious, but only weak. Men who were glad to have him at their houses for dinner, and who felt perfectly safe in allowing him to enter at will into their families, drew the line at golf. When a twosome or a foursome was proposed, and Watkins indicated a desire to be of the number, the others hastily hatched excuses, and the match dwindled down to him and the Colonel. So day after day he went the rounds alone, or accompanied by a faithful caddie who had grown callous. At first, quite a retinue of unemployed caddies attended Watkins from hole to hole, with delighted awe at his vocabulary. They furnished a breakwater to his overflow of irritation, until even they gave way, and were seen no more in his neighbourhood. When it is stated that Watkins really was not conscious of the frequency and liberality with which he swore, it will readily be understood that reformation, in such a case, is hard to accomplish.

It was one afternoon early in November, and he was playing, as usual, by himself, and indulging in his customary soliloquies. A light snow had fallen the day before, and, drifting over the field, had collected in little patches. Watkins had been unwise enough to attempt to use white balls, with the consequences that already he had lost three, the spots of snow making it a difficult matter to locate the spheres.

His feelings gained in warmth, until the expressions to which he was giving vent actually melted nearby bits of snow. Then, when he lost his fourth, and last, ball, his words were shocking—simply shocking. Back and forth through the fair green he walked, covering the area within which he judged the ball must be lying. With his eyes feverishly taking in every inch of the ground he marked his rapid nervous steps by fervent objurgations. But it was a choice of finding the ball or of going home, and the combativeness that made of Watkins an acute businessman and an ardent golfer kept him to his quest long after weaker natures would have abandoned the task.

Suddenly and unexpectedly, he found a ball. It was not the one he had lost, for when he picked it up it gleamed with frost. Also, it bore a strange trademark, almost effaced, but still showing a few letters. As in the instance of many a golf ball lost and found, it was lying right in the open, so that when once seen one marvelled how it ever could have been overlooked. With a sense of great satisfaction, Watkins stooped and grasped it. Probably there is no purer pleasure in life than that of finding a golf ball that someone else has lost. By the laws of golf, finders are keepers.

Watkins's irritation vanished instantly. This was a good ball—better than the one for which he had been seeking. It was about where his own lost ball should be, and therefore he was content to play it from where it had been lying. With care he replaced it, and drawing a mashie from his bag, he prepared to atone for wasted time by an accurate approach shot.

"Now, [blank] you!" he ejaculated, menacingly, taking his stance, and, in a double significance, addressing the ball. Forthwith he did a thing that had not happened to him in months—he swung right over the globe, missing it entirely! His ephemeral good nature yielded to a startling reaction. A distinct door of brimstone pervaded the atmosphere.

When, at the next attempt, he ploughed into the turf; his language increased in force; and when in quick succession he committed every fault that attends the merest novice, and ended by slicing the ball far off the course, into the rough, he was thoroughly aroused. Never in all his previous playing had he perpetrated such a series of outrageous blunders. And yet as he walked to where his last stroke had landed the sphere, his frenzy settled to a stony calmness. His long, angry strides shortened, becoming less pronounced as he proceeded, and when he reached the ball, lying there with its cunning, shining countenance staring up at him so blandly, he did not even stamp on it. With exceeding coolness he surveyed the position—a bad one, amid the ragweed—and deliberately selected a club from his bag. He forced a pleasant smile to his face, and whistled a popular melody.

"My little man," he mildly said to the ball, "we'll have to get out of here."

Thus speaking, he carefully measured with his eye the distance, paid close attention to his stance, and drawing a deep breath, struck, civilly, skilfully. It was a beautiful stroke. The ball sailed over the weeds, down the course, and stopped on the green about four feet from the hole. Watkins smiled—naturally this time. Then picking up his bag, he walked after the ball.

But in the glow of pleasure that crowned his victory, he forgot the courtesy due to the conquered. As, putter in hand, he arrived on the green, he hailed his antagonist with the open menace of old.

"Now, [blank] you!" he said.

If there was one department of golf in which, above all others, Watkins thought he excelled, it was putting. This was an easy putt, too. The hole was a scant four feet away, and the green level as a billiard table. Nevertheless, either Watkins was overconfident, or he took his eye off the ball, or—or something else must have happened; at any rate, he missed the hole by eight inches, and went five feet on the farther side.

Watkins swore plainly, unreservedly, and vehemently.

Again he over putted, this time jumping the hole, and going on and on until it seemed to him that the ball had legs. Already he had run up to twelve strokes. Was he never going to make that hole?

Another fruitless essay, but Watkins, by a strong effort, repressed his righteous indignation. For the second time this day he calmly approached the ball, calmly prepared for the stroke, calmly and sweetly smiled, and with the gentle admonition, "We'll see if we cannot do better this trial," putted.

The ball tumbled snugly into the hole. Watkins, picking out the sphere, examined it with attention. Solemnly he teed it, and not having uttered another syllable, by a clean drive he sent it flying for hole No. 5.

During ensuing days the club members noted a curious irregularity in Watkins' scores. He did the course in as low as 43 (it was a nine-hole course), and in as high as 80. The one surprised his friends as much as did the other, for Watkins was not a crack, even when on his game, but then neither was he in the duffer class.

He grew extremely irritable, and his uncertain temper was noted with sorrowful surprise by his business associates. It is true that he did not swear nearly so much on the links, but as an offset, a number of persons averred that they had heard him using unparliamentary language when not on the course. The truly moral pointed to Watkins as an example of the insidious nature of profanity.

"He acquired the dreadful habit playing golf, and now he can't control himself at all," they asserted, pityingly.

The day of the last club match of the season was at hand. When the qualifying scores were posted Watkins' read: 72, 41, 53. The handicap committee was, in consequence, somewhat puzzled as to the proper handicap, if any—for this affair was a free-for-all, with generous allowances.

What could be done with a man who might make the course in 40, but who was just as likely to make it in 70 or 80?

When Watkins teed his ball, a little murmur of astonishment swept through the spectators. It was not a new clean ball, such as is usually preferred for important contests. It was a wretched, dissipated-looking globe, scarred by cuts, and mottled with black where the paint had been worn from the gutta-percha.

"Watkins must have manufactured it from an old rubber boot," sarcastically observed De Lancey.

But no one ventured to suggest or to protest. Watkins, wholly indifferent to public opinion, drove for two hundred yards, and left the crowd behind.

Through the testimony of his caddie, we know that Watkins started by playing the best amateur game ever seen on the course of the Upland Golf Club. He made hole No. 1 in 5, No. 2 in 4, Nos. 3 and 4 in 6, No. 5 in 3, No. 6 in 4—so far only 1 over—and then—and then—ah!

The caddie has laid especial stress on the wonderful, unprecedented calmness that characterized Watkins' progress up to this spot. The caddie, in a measure, blames himself for what then occurred. He says that by mistake he handed his employer a driving mashie when the cleik had been called for, and that while Watkins did not offer any objection, and the lie was adapted for either club, nevertheless the stroke proved a miserable effort—quite the worse Watkins had made that day.

Although everyone is at times liable to make a misplay, Watkins instantly turned on the luckless caddie and swept him with a storm of upbraiding. The caddie would not repeat the language but is of the firm conviction that it more than made up his employer's general average of things one would have wished to have left unsaid.

The ball had moved only a few feet, and now was in a nasty bunker. Watkins stood over the place and swore vigorously. He appeared to revel in his freedom. Finally he endeavoured to extricate the ball. Four times he struck at it and succeeded only in forcing it deeper into the sand. At last he lifted it partly out. The bunker was at the bottom of a slight declivity, and halfway up this the ball stopped. The lie was fair, although rather ticklish. Watkins, blaspheming joyously at this deliverance, hastened to the ball, anxious to regain lost ground. Whether he jarred the earth or not the caddie does not know; but he does know that just as Watkins swung his club the ball rolled—back, ten yards, plump into the bunker.

The caddie says that the stillness was awful. Watkins jaws came together with a snap, the masseter muscles—the bunch at the point of the jaw on either cheek—knotting and hardening convulsively. He dropped his club, walked resolutely to the bunker, grabbed the ball, continued to the edge of the pond near hole No. 6, and with great force threw the offender far into the midst of the waters.

Of course Watkins lost the medal—he had lost it, anyway, by the disaster at the bunker. But from that day to this no one has heard him swear. For some weeks, it is true, he was very irritable, as is anybody who suddenly abandons smoking or drinking or other stimulating habit, but that quickly passed. Unruffled, imperturbable, serene, he does his daily round, a moral high-minded gentleman whose fiercest indignation never goes beyond a "tut! tut!" And it is excellent golf that he plays. Indeed Watkins is now rated as a crack, and plays at two behind scratch.

Today, somewhere in the mud in the centre of hole No. 6 pond, and despised by the resident community of other whiter balls exported there by the clubs of various players, is this potted, disfigured old rounder of a ball, secure from the heat and cold, bunker and whins, brassie, cleik, and niblick. Perhaps in its odd little heart it is conscious that it has not lived in vain.

Edwin Legrand Sabin (1870–1952) was an American historian and author who wrote a handful of science fiction short stories and a Lost World novel called "The City of the Sun". He is best known as the author of studiously researched adventure books for boys.

Commentary

The author of this golfing story, Edwin Sabin, published a collection of links-based tales. (For a short time, as the nineteenth century turned into the twentieth, readers could not get enough of golfing stories.) Sabin wrote this particular tale as a light, humorous, throwaway fantasy. But might the events he pictured one day come to pass? Not so much on the golf course, necessarily, but in our mundane, quotidian, everyday activities? Indeed, perhaps we already live in a world of 'supersensitive golf balls'—or similar equipment?

Those of us lucky enough to live in the developed world find ourselves in an age of automation. Machines now perform tasks we once thought humans alone could carry out. We can attribute much of this development to advances in artificial intelligence (AI) and machine learning (ML). (Writers often use the initialisms AI and ML interchangeably, so I will do the same here, but bear in mind the distinctions between the two terms. AI refers to technologies that can mimic human cognitive functions. ML is a subset of AI that enables a system to learn from experience. Whether either approach has anything to do with how humans think remains unclear.)

Some people already outsource a variety of chores to AI. Their intelligent fridges track food levels and use-by dates, and order replenishments as needed. Their smart dishwashers check detergent levels and re-supply themselves when necessary. Their robot vacuum cleaners keep floors free from dust. (One can imagine a range of future possibilities here. For example, what about an augmented-reality wardrobe? It helps you choose outfits based on your schedule, the weather, and fashion trends. It tracks when your clothes need dry cleaning or repair and schedules a pick-up and drop-off. It connects to a smart laundry hamper to sort dirty clothes by colour and fabric and then sets the appropriate settings on the washing machine. Or does the home of the typical billionaire already have such magical wardrobes?)

The influence of ML and AI makes itself felt outside of the home, too, and in various ways. Consider healthcare. Data from wearables can provide patients and doctors with advance notice of a medical emergency. Or consider fundamental science. The AI program AlphaFold can predict a protein's shape in three dimensions from its linear sequence, with potentially transformational applications in biology. In physics, some string theorists hope a similar approach might help them understand how extra dimensions compactify (see Chap. 1). Or consider agriculture. AI-powered systems can use data from sensors to optimize crop yields, reduce water usage, and detect plant diseases. Or consider accessibility, a field in which AI can power assistive technologies. Computer vision algorithms help the visually impaired navigate their surroundings. Natural language processors help lower the barriers to communication for deaf people. AI-powered prosthetics and exoskeletons help those with mobility impairments. No-one should disparage these developments. But those same techniques could infiltrate every aspect of life. Do we want this?

Consider media streaming services. They record what you watch and what you listen to, then use that history to suggest similar items. Such services could go further and study in real time your reactions, both conscious and subconscious, to what you watch and listen to. Then, with laser-like precision, they could suggest new shows and music to appeal to your preferences.

Certain fast food restaurants hope facial recognition systems will 'improve the customer experience'. How do they work? Well, suppose you frequent a particular restaurant. The system installed at your favourite venue gets to understand your food preferences. It recognises you when you step through the door, and tailors the menu items to your taste.

Some researchers have even proposed that AI could remove the hassle of having to exercise your right to vote. All citizens would have an AI representative, an avatar, that learns their outlook about various issues. The avatars would then deliberate on behalf of people, expressing opinions on issues from the local level up to the national. An AI that votes for you! (In some ways this reminds me of Asimov's 1955 story "Franchise". Asimov postulates that Multivac, a powerful computer, can predict the outcome of an election from a single voter.)

When we contemplate all these advances, Sabin's notion of a supersensitive golf ball seems not so outlandish. AI systems already tweak the world to make life more convenient, more frictionless, more accommodating. And when they fail, we feel as irritated as Watkins did when playing golf on the links.

These proliferating AI/ML systems undoubtedly provide convenience. But convenience hides costs.

If your streaming services push you a non-stop diet of TV soaps, you might never discover your hidden passion for foreign-language films. If your music services play middle-of-the-road pop music all day, you might never encounter Bach, Beethoven, and the Beatles. If your restaurant menu displays nothing but 'hamburger', you might never discover your latent love of crispy cauliflower. A world of smooth convenience, certainly, but narrow to the point of tedium. Worse than that, companies often train their AI systems on biased data and run them with biased algorithms. This can lead to discriminatory treatment. Facial recognition systems, we know, have issues with racial and gender bias. A nuisance if your fast food restaurant provides a menu filled with food you can't stomach; the implications become serious if law enforcement officers target you on the basis of false information.

And if we outsource the right to franchise to our avatars, will that lead to a healthier democracy? Perhaps it might increase voter involvement, which in some countries is lamentable. In 2020, in the USA, one in three eligible voters did not trouble themselves to cast a ballot in the presidential election. In 2021, in England, two in three on the electoral roll did not inconvenience themselves to vote in local elections. We should welcome any measure that increases participation in democracy, but might avatar democracy lead people to ignore politics entirely? (And would our avatars, upon comprehending the boundless laziness of their human counterparts, bother to deliberate in an AI parliament?) Avatar democracy might even go beyond the replacement of voters from politics and deliver the ostensibly attractive proposition of removing politicians from politics. But would the relief of not having to worry about politicians outweigh the risks? We would, after all, have to trust our avatars—just as we have to trust our politicians today. And power would flow to the avatar-makers. Could we trust those who produce, promote, or sell avatars?

In his 1947 story "With folded hands", Jack Williamson hinted at an even more fundamental concern with this technology. In his story, a new engineering company programs its robots to protect humans from hazards. The robots succeed in their mission. They succeed too well. People soon feel frustrated because the overprotective robots allow humans to do little, except sit—with folded hands. The scenario Williamson sketched might seem outlandish, but we can already see signs of a related development. The technophilic SF authors of earlier generations once promised us that AI could take over the jobs of drudgery, leaving people free to undertake tasks requiring imagination and originality. Instead, with the rise of generative artificial intelligence, it seems the reverse might come to pass. AI enjoys the creative work (writing screenplays, making music, creating art). Humans suffer the grind.

As with every technology humans have ever invented, so with AI. If we deploy it with wisdom, we will reap benefits. If we use it with folly, we will suffer. The human race could enjoy that SF dream of living lives rich in creativity. Or we could sleepwalk into a future of passivity, lack of agency, and reduced serendipity. Imagine a life in which you never get to feel that moment of unexpected delight that comes from venturing outside your comfort zone and encountering something novel. Any technology that optimises experiences based on our past preferences risks shielding us from those encounters that lead to new perspectives and personal growth.

Sabin here wrote a gentle fantasy. Stories by Asimov, Williamson, and similar writers had a harder edge. These latter authors used science fiction to carry out thought experiments, to explore the implications of living in a world in which inanimate objects exhibit agency. Science fiction delivered warnings, which we still have time to heed. We still have the chance to put in place proper oversight and accountability for this technology. I wonder whether we have the wisdom to do so?

Notes and Further Reading

The AI program AlphaFold—Machine learning is a subset of artificial intelligence technologies. The deep learning approach is a subset of machine learning methods, in which the learning network possesses multiple layers. The approach can discern patterns in vast amounts of data. One of the best known deep learning programs is AlphaFold by DeepMind. Researchers trained AlphaFold on thousands of known protein structures. The layered network was then able to predict the shape of other proteins based on their amino acid sequence. AlphaFold's achievement represents a stunning breakthrough because proteins perform crucial functions in our bodies. To carry out those functions, though, proteins must fold into specific three-dimensional shapes. Until now, there was no way of knowing how a particular sequence of amino acids would fold in three dimension. If biologists can understand how proteins fold, they can better understand the mechanisms behind diseases and they can design new medicines. See, for example, Eisenstein (2021). In 2024, scientists unveiled their AlphaFold 3 model, which is significantly more accurate than the original AlphaFold; see Abramson et al. (2024). This is a fast-moving field!

string theorists hope that a similar approach—We saw in Chap. 1 how string theorists, by positing the existence of extra space dimensions, have developed a theory that treats both gravity and the quantum world within a single, consistent framework. For the theory to work there must be six extra space dimensions. And to explain why we don't observe those extra dimensions, the string theorists argue those dimensions are small. The dimensions 'compactify', by curling up into a complex shape called a Calabi–Yau manifold. The problem for string theorists is that there is an astronomically large number of Calabi–Yau manifolds, and each different manifold would give rise to different physics. To describe our universe we would have to pick the right Calabi–Yau manifold from the vast number of possibilities. Some physicists hope that deep learning approaches might provide insight into compactification. For a non-technical introduction see, for example, Wood (2024).

fast food restaurants hope facial recognition systems—This is another fast-moving field. For an article describing the situation as of April 2024, see Applin (2024). For information about the possibility of discrimination in these systems, see for example Najibi (2020).

AI could remove the hassle of having to exercise one's right to vote—A TED Talk by César Hidalgo provides an accessible introduction to the idea of avatar democracy; see Hidalgo (2018). For more on how AI might interact with our notions of democracy, and of the ethical questions surrounding this, see Jungherr (2023) and references therein. For details on English local election turnout, see UK Parliament (2023); for details on the US Presidential election turnout, see US Census Bureau (2021).

Asimov's 1955 story "Franchise"—In his robot stories, the first of which was published in 1940, Asimov considered many of the questions that society is only now begin to grapple with. We had plenty of warning! The Multivac stories are about computers, rather than robots, but involve much the same sorts of ethical questions. The story "Franchise" was first published in the August 1955 issue of the magazine *If, Worlds of Science Fiction*. It can be found in the collection *Earth is Room Enough* (Asimov 1957).

his 1947 story "With folded hands"—Jack Williamson said the theme of his story "With folded hands" was that the perfect machine would prove to be perfectly destructive. The novella was first published in the July 1947 issue of *Astounding Science Fiction*, and can be found in the collection *The Science Fiction Hall of Fame, Volume Two* (Bova 1973).

References

Abramson, J., et al.: Accurate structure prediction of biomolecular interactions with AlphaFold 3. Nature. (2024). https://doi.org/10.1038/s41586-024-07487-w

Applin, S.A.: How fast food is becoming a new surveillance ground. FastCompany. www.fastcompany.com/91087484/how-fast-food-is-becoming-a-new-surveillance-ground (2024)

Asimov, I.: Earth is Room Enough. Doubleday, New York (1957)

Bova, B.: The Science Fiction Hall of Fame, vol. 2. Doubleday, New York (1973)

Eisenstein, M.: Artificial intelligence powers protein-folding predictions. Nature. **599**, 706–708. www.nature.com/articles/d41586-021-03499-y (2021)

Hidalgo, C.: A bold idea to replace politicians. TED Talk. www.ted.com/talks/cesar_hidalgo_a_bold_idea_to_replace_politicians (2018)

Jungherr, A.: Artificial intelligence and democracy: a conceptual framework. Social Media+Society. **9**(3) (2023). https://doi.org/10.1177/20563051231186353

Najibi, A.: Racial discrimination in face recognition technology. Science Policy and Social Justice. Harvard Graduate School of Arts and Sciences, Special Edition. https://sitn.hms.harvard.edu/special-edition-science-policy-and-social-justice/ (2020)

UK Parliament: Turnout at Elections. House of Commons Library. https://commonslibrary.parliament.uk/research-briefings/cbp-8060/ (2023)

US Census Bureau: 2020 Presidential Election Voting and Registration Tables Now Available. www.census.gov/newsroom/press-releases/2021/2020-presidential-election-voting-and-registration-tables-now-available.html (2021)

Wood, C.: AI starts to sift through string theory's near-endless possibilities. Quanta Magazine. www.quantamagazine.org/ai-starts-to-sift-through-string-theorys-near-endless-possibilities-20240423/#0. (2024)

6

The Science of Persuasion

Filboid Studge, the Story of a Mouse that Helped (Saki)

"I want to marry your daughter," said Mark Spayley with faltering eagerness. "I am only an artist with an income of two hundred a year, and she is the daughter of an enormously wealthy man, so I suppose you will think my offer a piece of presumption."

Duncan Dullamy, the great company inflator, showed no outward sign of displeasure. As a matter of fact, he was secretly relieved at the prospect of finding even a two-hundred-a-year husband for his daughter Leonore. A crisis was rapidly rushing upon him, from which he knew he would emerge with neither money nor credit; all his recent ventures had fallen flat, and flattest of all had gone the wonderful new breakfast food, Pipenta, on the advertisement of which he had sunk such huge sums. It could scarcely be called a drug in the market; people bought drugs, but no one bought Pipenta.

"Would you marry Leonore if she were a poor man's daughter?" asked the man of phantom wealth.

"Yes," said Mark, wisely avoiding the error of over-protestation. And to his astonishment Leonore's father not only gave his consent, but suggested a fairly early date for the wedding.

"I wish I could show my gratitude in some way," said Mark with genuine emotion. "I'm afraid it's rather like the mouse proposing to help the lion."

"Get people to buy that beastly muck," said Dullamy, nodding savagely at a poster of the despised Pipenta, "and you'll have done more than any of my agents have been able to accomplish."

"It wants a better name," said Mark reflectively, "and something distinctive in the poster line. Anyway, I'll have a shot at it."

Three weeks later the world was advised of the coming of a new breakfast food, heralded under the resounding name of "Filboid Studge". Spayley put forth no pictures of massive babies springing up with fungus-like rapidity under its forcing influence, or of representatives of the leading nations of the world scrambling with fatuous eagerness for its possession. One huge sombre poster depicted the Damned in Hell suffering a new torment from their inability to get at the Filboid Studge which elegant young fiends held in transparent bowls just beyond their reach. The scene was rendered even more gruesome by a subtle suggestion of the features of leading men and women of the day in the portrayal of the Lost Souls; prominent individuals of both political parties, Society hostesses, well-known dramatic authors and novelists, and distinguished aeroplanists were dimly recognizable in that doomed throng; noted lights of the musical-comedy stage flickered wanly in the shades of the Inferno, smiling still from force of habit, but with the fearsome smiling rage of baffled effort. The poster bore no fulsome allusions to the merits of the new breakfast food, but a single grim statement ran in bold letters along its base: "They cannot buy it now".

Spayley had grasped the fact that people will do things from a sense of duty which they would never attempt as a pleasure. There are thousands of respectable middle-class men who, if you found them unexpectedly in a Turkish bath, would explain in all sincerity that a doctor had ordered them to take Turkish baths; if you told them in return that you went there because you liked it, they would stare in pained wonder at the frivolity of your motive. In the same way, whenever a massacre of Armenians is reported from Asia Minor, every one assumes that it has been carried out "under orders" from somewhere or another; no one seems to think that there are people who might like to kill their neighbours now and then.

And so it was with the new breakfast food. No one would have eaten Filboid Studge as a pleasure, but the grim austerity of its advertisement drove housewives in shoals to the grocers' shops to clamour for an immediate supply. In small kitchens solemn pig-tailed daughters helped depressed mothers to perform the primitive ritual of its preparation. On the breakfast-tables of cheerless parlours it was partaken of in silence. Once the womenfolk discovered that it was thoroughly unpalatable, their zeal in forcing it on their households knew no bounds. "You haven't eaten your Filboid Studge!" would be screamed at the appetiteless clerk as he turned wearily from the breakfast-table, and his evening meal would be prefaced by a warmed-up mess which would be explained as "your Filboid Studge that you didn't eat this morning." Those strange fanatics who ostentatiously mortify themselves, inwardly and outwardly, with health biscuits and health garments, battened aggressively on the new food. Earnest spectacled young men devoured it on the steps of the National Liberal Club. A bishop who did not believe in a future state preached against the poster, and a peer's daughter died from eating too much of the compound. A further advertisement was obtained when an infantry regiment mutinied and shot its officers rather than eat the nauseous mess; fortunately, Lord Birrell of Blatherstone, who was War Minister at the moment, saved the situation by his happy epigram, that "Discipline to be effective must be optional."

Filboid Studge had become a household word, but Dullamy wisely realized that it was not necessarily the last word in breakfast dietary; its supremacy would be challenged as soon as some yet more unpalatable food should be put on the market. There might even be a reaction in favour of something tasty and appetizing, and the Puritan austerity of the moment might be banished from domestic cookery. At an opportune moment, therefore, he sold out his interests in the article which had brought him in colossal wealth at a critical juncture, and placed his financial reputation beyond the reach of cavil. As for Leonore, who was now an heiress on a far greater scale than ever before, he naturally found her something a vast deal higher in the husband market than a two-hundred-a-year poster designer. Mark Spayley, the brainmouse who had helped the financial lion with such untoward effect, was left to curse the day he produced the wonder-working poster.

"After all," said Clovis, meeting him shortly afterwards at his club, "you have this doubtful consolation, that 'tis not in mortals to countermand success."

Hector Hugh Munro (1870–1916), an English author, is better known under his pen name: Saki. When Munro was a child, a cow charged at his mother Mary; she miscarried, and died soon after from shock. Perhaps this explains his fear of the natural world, which found expression in a torrent of perfectly crafted, hard-edged, blackly comic short stories. He enlisted in 1914, and died two years later in the Battle of the Somme.

Commentary

Gentle reader, I see you've made it almost halfway through the book. Congratulations! You're a person of discernment. And now I've got something truly special to share with you. Something that could change your life forever. Tell me, have you ever found yourself nodding along to a sales pitch, only to realise later you've been duped? Ever felt tired of falling for underhanded advertising tactics? Want to arm yourself with the knowledge to fight back? Well, do I have the chapter for you! Packed with the latest research, this chapter unveils the psychological secrets behind why you say 'yes' when you want to scream 'no'. And the best part? This is not just any research. This is **cutting-edge**, **groundbreaking** research that will leave you wondering how you survived without it. So what are you waiting for? You've already admitted you need this information. Just sit back, relax, and …

Well, you get the idea.

Saki wrote his satire on the nature of advertising in 1910, just two years after a journalist had dubbed Thomas J. Barratt the 'father of modern advertising'. Barratt worked in London. Other countries had well established advertising operations, but Barratt became the world's first 'brand manager'. He had married the eldest daughter of Francis Pears, head of the soap company A. & F. Pears, and entered the firm in 1865. There, he instituted a systematic method of advertising the soap, an approach that combined distinctive images with catchy slogans. See Fig. 6.1. In comparison with today's techniques, Barratt's 'modern' advertising methods seem primitive. For a start, in Saki's time print formed the basis of almost all paid advertising: knowledge of Filboid Studge, as with real consumer products, reached the Edwardian public through the medium of posters. Today, as print advertising declines total advertising spend rises. In 2023, the worldwide advertising spend exceeded $1 trillion. Ads now bombard consumers on diverse platforms—radio, television, cinema, and increasingly varied digital outlets. More intriguingly, though, advertisers today understand consumer behaviour in ways Thomas J. Barratt could not have dreamed.

Saki's story rests on an intriguing point of psychology: people will often buy something not through desire or need but because they feel they *should*. Since his story appeared, the field of behavioural science has burgeoned. Behavioural science draws on anthropology, behavioural economics, psychology, and sociology to study human action. It looks at how people behave and how they reach decisions. Professionals in various fields employ lessons from behavioural science to help them deliver desirable outcomes. For example,

Fig. 6.1 Left: A 1900 poster for Pears soap ("Matchless for the Complexion") created for Barratt. Right: a 1902 Harry Furniss cartoon satirising advertising testimonials. (Public domain)

officials responsible for public health can use the findings of behavioural science to encourage people to make healthy food choices. Those involved in environmental health can use the research to help people choose environmentally friendly options over unsustainable ones. Officials who wish to prevent fraud can use the results to help people interact with computers in a more secure way. But those same research findings from behavioural science provide advertising executives and marketing teams with powerful tools. Advertisers know how best to sell us stuff—either physical goods or services—regardless of whether we want or need them.

One example of an advertising tactic, rooted in a finding from behavioural science that Saki would surely recognise, is the 'scarcity principle'. When goods or services are in short supply, people want them more—as if customers equate attractiveness with limited availability. Tell your clients that this is 'the last one', or that 'the special deal ends soon', then you increase your chances of getting the sale. People have a fear of missing out. Which is why when you go on a travel booking site, for example, you will almost certainly come across a statement such as: "In high demand—only two rooms left". Or "Availability is low on these dates—lock in a great price before it's gone". Or the classic "Offer ends tonight".

Advertisers can deploy various other tactics of persuasion, all of which have had their effectiveness confirmed by behavioural science. Modern technologies have amplified the potency of those tactics. The ubiquity of screens exposes us to a constant stream of marketing messages, as if the advertising posters of Saki's day follow us around. The way we interact with those messages generates vast amounts of data, so advertising firms now hold granular information on each of us. Each customer becomes a data point in an abstract higher-dimensional space—a collection of numbers representing demographics, personality, interests, and much more—which machine learning techniques, as we saw in Chap. 5, can analyse in exquisite detail. Based on this intimate profiling, firms can customize an ad to play on each individual consumer's fears, values, and sense of identity. They can manipulate emotional triggers unique to each person, with messages honed through focus groups and validated by psychological research. Concerned about the environment? You'll be the target of ads for 'green' products. The ads will use activist messaging and imagery, and they'll appeal to your self-image as an eco-friendly individual. Feel depressed or anxious? You'll see ads for pharmaceuticals related to mental health. Interested in fitness? You'll be the recipient of ads for sportswear, food supplements, and slimming apps that take account of factors such as age, gender, race, and location.

With online advertising now central to the tech industry's profits, companies have incentives to continue perfecting their emotionally manipulative targeting. Unless legislators put in place regulation to curb the most ethically dubious practices of digital advertisers, the trend towards hyper-personalised advertising, with its capacity to shape choices both as consumers and as citizens in a democracy, might intensify. One can imagine how, as digital life becomes more immersive, ads will surround consumers in simulated environments. Biometric data will precisely track emotional responses to those ads and allow messages to be further honed. Saki's mordant take on advertising seems increasingly relevant as technology evolves.

The future will have aspects other than the digital, of course, and SF writers have a long history of exploring how advertisers could take advantage of various technological advances. Consider four of the most famous SF authors. H. G. Wells, in his novel *The Sleeper Awakes*, imagined a future of "great fleets of advertisement balloons and kites". Robert A. Heinlein, in his short novel *The Man Who Sold the Moon*, described how advertisers could use carbon black to draw ads on our satellite's surface. Arthur C. Clark, in his short story

"Watch this space", suggested something similar: a lunar-based ad for a carbonated soft drink (Coca-Cola, let's face it) made from a cloud of glowing sodium. And Isaac Asimov, in his short story "Buy Jupiter", told the humorous tale of how aliens acquired the rights to use Jupiter as a cosmic billboard. Perhaps advertising on that astronomical scale might one day come to pass. But I suspect Saki's more subtle take, which has its roots in human psychology, better reflects the reality of advertising in a digital world.

Notes and Further Reading

the worldwide adversing spend exceeded $1 trillion—Information on ad spending, broken down by country and by platform, can be found at Statista (n.d.).

the field of behavioural science has burgeoned—It is difficult to pinpoint the origin of behavioural science, in part because the term is difficult to define. The UN, which looks to behavioural science to guide its actions on climate change, vaccine rollouts, and other major challenges, defines it as "an evidence-based understanding of how people actually behave, make decisions and respond to programmes, policies, and incentives" (UN, n.d.). The field draws from multiple disciplines, but one particularly important element is behavioural economics. The Israeli–American psychologist Daniel Kahneman (1934–2024), who was awarded the 2002 Nobel Memorial Prize in Economic Sciences, was one of the pioneers of behavioural economics. His book *Thinking, Fast and Slow* (Kahneman 2011), is a clear account of the biases that affect human decision making. One of Kahneman's colleagues, the American economist Richard Thaler (1945–), used these insights to further refine the field of behavioural economics. Thaler was awarded the Nobel Prize in 2017. His book *Nudge* (Thaler and Sunstein 2008), which explores how organisations can help people make better choices, has been hugely influential: governments around the world have set up 'nudge units'. The same insights, however, can be used by less altruistic actors. Kuyer and Gordijn (2023) review the ethical issues with nudging.

various other tactics of persuasion—The American psychologist and behavioural scientist Robert Beno Cialdini (1945–) has studied how salespeople and fundraisers persuade people. He identified six key principles, including the scarcity principle, that sellers use to influence consumers. He describes these principles in his book *Influence: The Psychology of Persuasion* (Cialdini 1984). With the rise of social media influencers, Cialdini has added a seventh key idea: the unity principle. This is the notion that the more we identify with other people, the more we are influenced by them.

a long history of exploring how advertisers—The works mentioned here are by Asimov (1958), Clarke (1957), Heinlein (1950), and Wells (1910).

References

Asimov, I.: Buy Jupiter. Venture Science Fiction. May. (1958) Reprinted in *Buy Jupiter and Other Stories.* Doubleday, New York (1975)

Cialdini, R.B.: Influence: The Psychology of Persuasion. William Morrow, New York (1984)

Clarke, A.C.: Watch this space. Magazine of Fantasy and Science Fiction. February (1957)

Heinlein, R.A.: The Man Who Sold the Moon. Shasta, Chicago (1950)

Kahneman, D.: Thinking, Fast and Slow. Farrar, Straus and Giroux, New York (2011)

Kuyer, P., Gordijn, B.: Nudge in perspective: a systematic literature review on the ethical issues with nudging. Ration. Soc. 35(2), 191–230 (2023). https://doi.org/10.1177/10434631231155005

Statista: Advertising–worldwide. www.statista.com/outlook/amo/advertising/worldwide (n.d.)

Thaler, R.H., Sunstein, C.: Nudge: Improving Decisions About Health, Wealth, and Happiness. Yale University Press, New Haven, CT (2008)

UN: Secretary-General's Guidance Note on Behavioural Science. www.un.org/en/content/behaviouralscience/ (n.d.)

Wells, H.G.: The Sleeper Awakes. Harper, New York (1910)

7

The Future of Science Publishing

With the Eyes Shut (Edward Bellamy)

Railroad rides are naturally tiresome to persons who cannot read on the cars, and, being one of those unfortunates, I resigned myself, on taking my seat in the train, to several hours of tedium, alleviated only by such cat-naps as I might achieve. Partly on account of my infirmity, though more on account of a taste for rural quiet and retirement, my railroad journeys are few and far between. Strange as the statement may seem in days like these, it had actually been 5 years since I had been on an express train of a trunk line. Now, as every one knows, the improvements in the conveniences of the best equipped trains have in that period been very great, and for a considerable time I found myself amply entertained in taking note first of one ingenious device and then of another, and wondering what would come next. At the end of the first hour, however, I was pleased to find that I was growing comfortably drowsy, and proceeded to compose myself for a nap, which I hoped might last to my destination.

Presently I was touched on the shoulder, and a train boy asked me if I would not like something to read. I replied, rather petulantly, that I could not read on the cars, and only wanted to be let alone.

"Beg pardon, sir," the train boy replied, "but I'll give you a book you can read with your eyes shut. Guess you haven't taken this line lately," he added, as I looked up offended at what seemed impertinence. "We've been furnishing the new-fashioned phonographed books and magazines on this train for six months now, and passengers have got so they won't have anything else."

Probably this piece of information ought to have astonished me more than it did, but I had read enough about the wonders of the phonograph to be prepared in a vague sort of way for almost anything which might be related of it, and for the rest, after the air-brakes, the steam heat, the electric lights and annunciators, the vestibuled cars, and other delightful novelties I had just been admiring, almost anything seemed likely in the way of railway conveniences. Accordingly, when the boy proceeded to rattle off a list of the latest novels, I stopped him with the name of one which I had heard favourable mention of, and told him I would try that.

He was good enough to commend my choice. "That's a good one," he said. "It's all the rage. Half the train's on it this trip. Where'll you begin?"

"Where? Why, at the beginning. Where else?" I replied.

"All right. Didn't know but you might have partly read it. Put you on at any chapter or page, you know. Put you on at first chapter with next batch in five minutes, soon as the batch that's on now gets through."

He unlocked a little box at the side of my seat, collected the price of three hours' reading at five cents an hour, and went on down the aisle. Presently I heard the tinkle of a bell from the box which he had unlocked. Following the example of others around me, I took from it a sort of two-pronged fork with the tines spread in the similitude of a chicken's wishbone. This contrivance, which was attached to the side of the car by a cord, I proceeded to apply to my ears, as I saw the others doing.

For the next three hours I scarcely altered my position, so completely was I enthralled by my novel experience. Few persons can fail to have made the observation that if the tones of the human voice did not have a charm for us in themselves apart from the ideas they convey, conversation to a great extent would soon be given up, so little is the real intellectual interest of the topics with which it is chiefly concerned. When, then, the sympathetic influence of the voice is lent to the enhancement of matter of high intrinsic interest, it is not strange that the attention should be enchained. A good story is highly entertaining even when we have to get at it by the roundabout means of spelling out the signs that stand for the words, and imagining them uttered, and then imagining what they would mean if uttered. What, then, shall be said of the delight of sitting at one's ease, with closed eyes, listening to the same story poured into one's ears in the strong, sweet, musical tones of a perfect mistress of the art of story-telling, and of the expression and excitation by means of the voice of every emotion?

When, at the conclusion of the story, the train boy came to lock up the box, I could not refrain from expressing my satisfaction in strong terms. In reply he volunteered the information that next month the cars for day trips on that line would be further fitted up with phonographic guide-books of the country the train passed through, so connected by clock-work with the running gear of the cars that the guide-book would call attention to every object in the landscape, and furnish the pertinent information—statistical, topographical, biographical, historical, romantic, or legendary, as it might be—just at the time the train had reached the most favourable point of view. It was believed that this arrangement (for which, as it would work automatically and require little attendance, being used or not, according to pleasure, by the passenger, there would be no charge) would do much to attract travel to the road. His explanation was interrupted by the announcement in loud, clear, and deliberate tones, which no one could have had any excuse for misunderstanding, that the train was now approaching the city of my destination. As I looked around in amazement to discover what manner of brakeman this might be whom I had understood, the train boy said, with a grin, "That's our new phonographic annunciator."

Hamage had written me that he would be at the station, but something had evidently prevented him from keeping the appointment, and as it was late, I went at once to a hotel and to bed. I was tired and slept heavily; once or twice I woke up, after dreaming there were people in my room talking to me, but quickly dropped off to sleep again. Finally I awoke, and did not so soon fall asleep. Presently I found myself sitting up in bed with half a dozen extraordinary sensations contending for right of way along my backbone. What had startled me was the voice of a young woman, who could not have been standing more than ten feet from my bed. If the tones of her voice were any guide, she was not only a young woman, but a very charming one.

"My dear sir," she had said, "you may possibly be interested in knowing that it now wants just a quarter of three."

For a few moments I thought—well, I will not undertake the impossible task of telling what extraordinary conjectures occurred to me by way of accounting for the presence of this young woman in my room before the true explanation of the matter occurred to me. For, of course, when my experience that afternoon on the train flashed through my mind, I guessed at once that the solution of the mystery was in all probability merely a phonographic device for announcing the hour. Nevertheless, so thrilling and lifelike in effect were the tones of the voice I had heard that I confess I had not the nerve to light the gas to investigate till I had indued my more essential garments. Of course I found no lady in the room, but only a clock. I had not particularly noticed it on going to bed, because it looked like any other clock, and so now it continued to behave until the hands pointed to three. Then, instead of leaving me to infer the time from the arbitrary symbolism of three strokes on a bell, the same voice which had before electrified me informed me, in tones which would have lent a charm to the driest of statistical details, what the hour was. I had never before been impressed with any particular interest attaching to the hour of three in the morning, but as I heard it announced in those low, rich, thrilling contralto tones, it appeared fairly to coruscate with previously latent suggestions of romance and poetry, which, if somewhat vague, were very pleasing. Turning out the gas that I might the more easily imagine the bewitching presence which the voice suggested, I went back to bed, and lay awake there until morning, enjoying the society of my bodiless companion and the delicious shock of her quarter-hourly remarks. To make the illusion more complete and the more unsuggestive of the mechanical explanation which I knew of course was the real one, the phrase in which the announcement of the hour was made was never twice the same.

Right was Solomon when he said that there was nothing new under the sun. Sardanapalus or Semiramis herself would not have been at all startled to hear a human voice proclaim the hour. The phonographic clock had but replaced the slave whose business, standing by the noiseless water-clock, it was to keep tale of the moments as they dropped, ages before they had been taught to tick.

In the morning, on descending, I went first to the clerk's office to inquire for letters, thinking Hamage, who knew I would go to that hotel if any, might have addressed me there. The clerk handed me a small oblong box. I suppose I stared at it in a rather helpless way, for presently he said: "I beg your pardon, but I see you are a stranger. If you will permit me, I will show you how to read your letter."

I gave him the box, from which he took a device of spindles and cylinders, and placed it deftly within another small box which stood on the desk. Attached to this was one of the two-pronged ear-trumpets I already knew the use of. As I placed it in position, the clerk touched a spring in the box, which set some sort of motor going, and at once the familiar tones of Dick Hamage's voice expressed his regret that an accident had prevented his meeting me the night before, and informed me that he would be at the hotel by the time I had breakfasted.

The letter ended, the obliging clerk removed the cylinders from the box on the desk, replaced them in that they had come in, and returned it to me.

"Isn't it rather tantalizing," said I, "to receive one of these letters when there is no little machine like this at hand to make it speak?"

"It doesn't often happen," replied the clerk, "that anybody is caught without his indispensable, or at least where he cannot borrow one."

"His indispensable!" I exclaimed: "What may that be?"

In reply the clerk directed my attention to a little box, not wholly unlike a case for a binocular glass, which, now that he spoke of it, I saw was carried, slung at the side, by every person in sight.

"We call it the indispensable because it is indispensable, as, no doubt, you will soon find for yourself."

In the breakfast-room a number of ladies and gentlemen were engaged as they sat at table in reading, or rather in listening to, their morning's correspondence. A greater or smaller pile of little boxes lay beside their plates, and one after another they took from each its cylinders, placed them in their indispensables, and held the latter to their ears. The expression of the face in reading is so largely affected by the necessary fixity of the eyes that intelligence is absorbed from the printed or written page with scarcely a change of countenance, which when communicated by the voice evokes a responsive play of features. I had never been struck so forcibly by this obvious reflection as I was in observing the expression of the faces of these people as they listened to their correspondents. Disappointment, pleased surprise, chagrin, disgust, indignation, and amusement were alternately so legible on their faces that it was perfectly easy for one to be sure in most cases what the tenor at least of the letter was. It occurred to me that while in the old time the pleasure of receiving letters had been so far balanced by this drudgery of writing them as to keep correspondence within some bounds, nothing less than freight trains could suffice for the mail service in these days, when to write was but to speak, and to listen was to read.

After I had given my order, the waiter brought a curious-looking oblong case, with an ear-trumpet attached, and, placing it before me, went away. I foresaw that I should have to ask a good many questions before I got through, and, if I did not mean to be a bore, I had best ask as few as necessary. I determined to find out what this trap was without assistance. The words "Daily Morning Herald" sufficiently indicated that it was a newspaper. I suspected that a certain big knob, if pushed, would set it going. But, for all I knew, it might start in the middle of the advertisements. I looked closer. There were a number of printed slips upon the face of the machine, arranged about a circle like the numbers on a dial. They were evidently the headings of news articles. In the middle of the circle was a little pointer, like the hand of a clock, moving on a pivot. I pushed this pointer around to a certain caption, and then, with the air of being perfectly familiar with the machine, I put the pronged trumpet to my ears and pressed the big knob. Precisely! It worked like a charm; so much like a charm, indeed, that I should certainly have allowed my breakfast to cool had I been obliged to choose between that and my newspaper. The inventor of the apparatus had, however, provided against so painful a dilemma by a simple attachment to the trumpet, which held it securely in position upon the shoulders behind the head, while the hands were left free for knife and fork. Having slyly noted the manner in which my Neighbors had effected the adjustments, I imitated their example with a careless air, and presently, like them, was absorbing physical and mental aliment simultaneously.

While I was thus delightfully engaged, I was not less delightfully interrupted by Hamage, who, having arrived at the hotel, and learned that I was in the breakfast-room, came in and sat down beside me. After telling him how much I admired the new sort of newspapers, I offered one criticism, which was that there seemed to be no way by which one could skip dull paragraphs or uninteresting details.

"The invention would, indeed, be very far from a success," he said, "if there were no such provision, but there is."

He made me put on the trumpet again, and, having set the machine going, told me to press on a certain knob, at first gently, afterward as hard as I pleased. I did so, and found that the effect of the "skipper," as he called the knob, was to quicken the utterance of the phonograph in proportion to the pressure to at least tenfold the usual rate of speed, while at any moment, if a word of interest caught the ear, the ordinary rate of delivery was resumed, and by another adjustment the machine could be made to go back and repeat as much as desired.

When I told Hamage of my experience of the night before with the talking clock in my room, he laughed uproariously.

"I am very glad you mentioned this just now," he said, when he had quieted himself. "We have a couple of hours before the train goes out to my place, and I'll take you through Orton's establishment, where they make a specialty of these talking clocks. I have a number of them in my house, and, as I don't want to have you scared to death in the night-watches, you had better get some notion of what clocks nowadays are expected to do."

Orton's, where we found ourselves half an hour later, proved to be a very extensive establishment, the firm making a specialty of horological novelties, and particularly of the new phonographic timepieces. The manager, who was a personal friend of Hamage's, and proved very obliging, said that the latter were fast driving the old-fashioned striking clocks out of use.

"And no wonder," he exclaimed; "the old-fashioned striker was an unmitigated nuisance. Let alone the brutality of announcing the hour to a refined household by four, eight, or ten rude bangs, without introduction or apology, this method of announcement was not even tolerably intelligible. Unless you happened to be attentive at the moment the din began, you could never be sure of your count of strokes so as to be positive whether it was eight, nine, ten, or eleven. As to the half and quarter strokes, they were wholly useless unless you chanced to know what was the last hour struck. And then, too, I should like to ask you why, in the name of common sense, it should take twelve times as long to tell you it is twelve o'clock as it does to tell you it is one."

The manager laughed as heartily as Hamage had done on learning of my scare of the night before.

"It was lucky for you," he said, "that the clock in your room happened to be a simple time announcer, otherwise you might easily have been startled half out of your wits." I became myself quite of the same opinion by the time he had shown us something of his assortment of clocks. The mere announcing of the hours and quarters of hours was the simplest of the functions of these wonderful and yet simple instruments. There were few of them which were not arranged to "improve the time", as the old-fashioned prayer-meeting phrase was. People's ideas differing widely as to what constitutes improvement of time, the clocks varied accordingly in the nature of the edification they provided. There were religious and sectarian clocks, moral clocks, philosophical clocks, free-thinking and infidel clocks, literary and poetical clocks, educational clocks, frivolous and bacchanalian clocks. In the religious clock department were to be found Catholic, Presbyterian, Methodist, Episcopal, and Baptist time-pieces, which, in connection with the announcement of the hour and quarter, repeated some tenet of the sect with a proof text. There were also Talmage clocks, and Spurgeon clocks, and Storrs clocks, and Brooks clocks, which respectively marked the flight of time by phrases taken from the sermons of these eminent divines, and repeated in precisely the voice and accents of the original delivery. In startling proximity to the religious department I was shown the skeptical clocks. So near were they, indeed, that when, as I stood there, the various time-pieces announced the hour of ten, the war of opinions that followed was calculated to unsettle the firmest convictions. The observations of an Ingersoll which stood near me were particularly startling. The effect of an actual wrangle was the greater from the fact that all these individual clocks were surmounted by effigies of the authors of the sentiments they repeated.

I was glad to escape from this turmoil to the calmer atmosphere of the philosophical and literary clock department. For persons with a taste for antique moralizing, the sayings of Plato, Epictetus, and Marcus Aurelius had here, so to speak, been set to time. Modern wisdom was represented by a row of clocks surmounted by the heads of famous maxim-makers, from Rochefoucauld to Josh Billings. As for the literary clocks, their number and variety were endless. All the great authors were represented. Of the Dickens clocks alone there were half a dozen, with selections from his greatest stories. When I suggested that, captivating as such clocks must be, one might in time grow weary of hearing the same sentiments reiterated, the manager pointed out that the phonographic cylinders were removable, and could be replaced by other sayings by the same author or on the same theme at any time. If one tired of an author altogether, he could have the head unscrewed from the top of the clock and that of some other celebrity substituted, with a brand-new repertory.

"I can imagine," I said, "that these talking clocks must be a great resource for invalids especially, and for those who cannot sleep at night. But, on the other hand, how is it when people want or need to sleep? Is not one of them quite too interesting a companion at such a time?"

"Those who are used to it," replied the manager, "are no more disturbed by the talking clock than we used to be by the striking clock. However, to avoid all possible inconvenience to invalids, this little lever is provided, which at a touch will throw the phonograph out of gear or back again. It is customary when we put a talking or singing clock into a bedroom to put in an electric connection, so that by pressing a button at the head of the bed a person, without raising the head from the pillow, can start or stop the phonographic gear, as well as ascertain the time, on the repeater principle as applied to watches."

Hamage now said that we had only time to catch the train, but our conductor insisted that we should stop to see a novelty of phonographic invention, which, although not exactly in their line, had been sent them for exhibition by the inventor. It was a device for meeting the criticism frequently made upon the churches of a lack of attention and cordiality in welcoming strangers. It was to be placed in the lobby of the church, and had an arm extending like a pump-handle. Any stranger on taking this and moving it up and down would be welcomed in the pastor's own voice, and continue to be welcomed as long as he kept up the motion. While this welcome would be limited to general remarks of regard and esteem, ample provision was made for strangers who desired to be more particularly inquired into. A number of small buttons on the front of the contrivance bore respectively the words, "Male", "Female", "Married", "Unmarried", "Widow", "Children", "No Children", etc., etc. By pressing the one of these buttons corresponding to his or her condition, the stranger would be addressed in terms probably quite as accurately adapted to his or her condition and needs as would be any inquiries a preoccupied clergyman would be likely to make under similar circumstances. I could readily see the necessity of some such substitute for the pastor, when I was informed that every prominent clergyman was now in the habit of supplying at least a dozen or two pulpits simultaneously, appearing by turns in one of them personally, and by phonograph in the others.

The inventor of the contrivance for welcoming strangers was, it appeared, applying the same idea to machines for discharging many other of the more perfunctory obligations of social intercourse. One being made for the convenience of the President of the United States at public receptions was provided with forty-two buttons for the different States, and others for the principal cities of the Union, so that a caller, by proper manipulation, might, while shaking a handle, be addressed in regard to his home interests with an exactness of information as remarkable as that of the traveling statesmen who rise from the gazetteer to astonish the inhabitants of Wayback Crossing with the precise figures of their town valuation and birth rate, while the engine is taking in water.

We had by this time spent so much time that on finally starting for the railroad station we had to walk quite briskly. As we were hurrying along the street, my attention was arrested by a musical sound, distinct though not loud, proceeding apparently from the indispensable which Hamage, like everybody else I had seen, wore at his side. Stopping abruptly, he stepped aside from the throng, and, lifting the indispensable quickly to his ear, touched something, and exclaiming, "Oh, yes, to be sure!" dropped the instrument to his side.

Then he said to me: "I am reminded that I promised my wife to bring home some story-books for the children when I was in town today. The store is only a few steps down the street." As we went along, he explained to me that nobody any longer pretended to charge his mind with the recollection of duties or engagements of any sort. Everybody depended upon his indispensable to remind him in time of all undertakings and responsibilities. This service it was able to render by virtue of a simple enough adjustment of a phonographic cylinder charged with the necessary word or phrase to the clockwork in the indispensable, so that at any time fixed upon in setting the arrangement an alarm would sound, and, the indispensable being raised to the ear, the phonograph would deliver its message, which at any subsequent time might be called up and repeated. To all persons charged with weighty responsibilities depending upon accuracy of memory for their correct discharge, this feature of the indispensable rendered it, according to Hamage, and indeed quite obviously, an indispensable truly. To the railroad engineer it served the purpose not only of a time-piece, for the works of the indispensable include a watch, but to its ever vigilant alarm he could intrust his running orders, and, while his mind was wholly concentrated upon present duties, rest secure that he would be reminded at just the proper time of trains which he must avoid and switches he must make. To the indispensable of the business man the reminder attachment was not less necessary. Provided with that, his notes need never go to protest through carelessness, nor, however absorbed, was he in danger of forgetting an appointment.

Thanks to these portable memories it was, moreover, now possible for a wife to intrust to her husband the most complex messages to the dressmaker. All she had to do was to whisper the communication into her husband's indispensable while he was at breakfast, and set the alarm at an hour when he would be in the city.

"And in like manner, I suppose," suggested I, "if she wishes him to return at a certain hour from the club or the lodge, she can depend on his indispensable to remind him of his domestic duties at the proper moment, and in terms and tones which will make the total repudiation of connubial allegiance the only alternative of obedience. It is a very clever invention, and I don't wonder that it is popular with the ladies; but does it not occur to you that the inventor, if a man, was slightly inconsiderate? The rule of the American wife has hitherto been a despotism which could be tempered by a bad memory. Apparently, it is to be no longer tempered at all."

Hamage laughed, but his mirth was evidently a little forced, and I inferred that the reflection I had suggested had called up certain reminiscences not wholly exhilarating. Being fortunate, however, in the possession of a mercurial temperament, he presently rallied, and continued his praises of the artificial memory provided by the indispensable. In spite of the criticism which I had made upon it, I confess I was not a little moved by his description of its advantages to absent-minded men, of whom I am chief. Think of the gain alike in serenity and force of intellect enjoyed by the man who sits down to work absolutely free from that accursed cloud on the mind of things he has got to remember to do, and can only avoid totally forgetting by wasting tenfold the time required finally to do them in making sure by frequent rehearsals that he has not forgotten them! The only way that one of these trivialities ever sticks to the mind is by wearing a sore spot in it which heals slowly. If a man does not forget it, it is for the same reason that he remembers a grain of sand in his eye. I am conscious that my own mind is full of cicatrices of remembered things, and long ere this it would have been peppered with them like a colander, had I not a good while ago, in self-defence, absolutely refused to be held accountable for forgetting anything not connected with my regular business.

While firmly believing my course in this matter to have been justifiable and necessary, I have not been insensible to the domestic odium which it has brought upon me, and could but welcome a device which promised to enable me to regain the esteem of my family while retaining the use of my mind for professional purposes.

As the most convenient conceivable receptacle of hasty memoranda of ideas and suggestions, the indispensable also most strongly commended itself to me as a man who lives by writing. How convenient when a flash of inspiration comes to one in the night-time, instead of taking cold and waking the family in order to save it for posterity, just to whisper it into the ear of an indispensable at one's bedside, and be able to know it in the morning for the rubbish such untimely conceptions usually are! How often, likewise, would such a machine save in all their first vividness suggestive fancies, anticipated details, and other notions worth preserving, which occur to one in the full flow of composition, but are irrelevant to what is at the moment in hand! I determined that I must have an indispensable.

The bookstore, when we arrived there, proved to be the most extraordinary sort of bookstore I had ever entered, there not being a book in it. Instead of books, the shelves and counters were occupied with rows of small boxes.

"Almost all books now, you see, are phonographed," said Hamage.

"The change seems to be a popular one," I said, "to judge by the crowd of book-buyers." For the counters were, indeed, thronged with customers as I had never seen those of a bookstore before.

"The people at those counters are not purchasers, but borrowers," Hamage replied; and then he explained that whereas the old-fashioned printed book, being handled by the reader, was damaged by use, and therefore had either to be purchased outright or borrowed at high rates of hire, the phonograph of a book being not handled, but merely revolved in a machine, was but little injured by use, and therefore phonographed books could be lent out for an infinitesimal price. Everybody had at home a phonograph box of standard size and adjustments, to which all phonographic cylinders were gauged. I suggested that the phonograph, at any rate, could scarcely have replaced picture-books. But here, it seemed, I was mistaken, for it appeared that illustrations were adapted to phonographed books by the simple plan of arranging them in a continuous panorama, which by a connecting gear was made to unroll behind the glass front of the phonograph case as the course of the narrative demanded.

"But, bless my soul!" I exclaimed, "everybody surely is not content to borrow their books? They must want to have books of their own, to keep in their libraries."

"Of course," said Hamage. "What I said about borrowing books applies only to current literature of the ephemeral sort. Everybody wants books of permanent value in his library. Over yonder is the department of the establishment set apart for book-buyers."

The counter which he indicated being less crowded than those of the borrowing department, I expressed a desire to examine some of the phonographed books. As we were waiting for attendance, I observed that some of the customers seemed very particular about their purchases, and insisted upon testing several phonographs bearing the same title before making a selection. As the phonographs seemed exact counterparts in appearance, I did not understand this till Hamage explained that differences as to style and quality of elocution left quite as great a range of choice in phonographed books as varieties in type, paper, and binding did in printed ones. This I presently found to be the case when the clerk, under Ham-age's direction, began waiting on me. In succession I tried half a dozen editions of Tennyson by as many different elocutionists, and by the time I had heard "Where Claribel low lieth" rendered by a soprano, a contralto, a bass, and a baritone, each with the full effect of its quality and the personal equation besides, I was quite ready to admit that selecting phonographed books for one's library was as much more difficult as it was incomparably more fascinating than suiting one's self with printed editions. Indeed, Hamage admitted that nowadays nobody with any taste for literature—if the word may for convenience be retained—thought of contenting himself with less than half a dozen renderings of the great poets and dramatists. "By the way," he said to the clerk, "won't you just let my friend try the Booth-Barrett Company's 'Othello'? It is, you understand," he added to me, "the exact phonographic reproduction of the play as actually rendered by the company."

Upon his suggestion, the attendant had taken down a phonograph case and placed it on the counter. The front was an imitation of a theatre with the curtain down. As I placed the transmitter to my ears, the clerk touched a spring and the curtain rolled up, displaying a perfect picture of the stage in the opening scene. Simultaneously the action of the play began, as if the pictured men upon the stage were talking. Here was no question of losing half that was said and guessing the rest. Not a word, not a syllable, not a whispered aside of the actors, was lost; and as the play proceeded the pictures changed, showing every important change of attitude on the part of the actors. Of course the figures, being pictures, did not move, but their presentation in so many successive attitudes presented the effect of movement, and made it quite possible to imagine that the voices in my ears were really theirs. I am exceedingly fond of the drama, but the amount of effort and physical inconvenience necessary to witness a play has rendered my indulgence in this pleasure infrequent. Others might not have agreed with me, but I confess that none of the ingenious applications of the phonograph which I had seen seemed to be so well worth while as this.

Hamage had left me to make his purchases, and found me on his return still sitting spellbound.

"Come, come," he said, laughing, "I have Shakespeare complete at home, and you shall sit up all night, if you choose, hearing plays. But come along now, I want to take you upstairs before we go."

He had several bundles. One, he told me, was a new novel for his wife, with some fairy stories for the children,—all, of course, phonographs. Besides, he had bought an indispensable for his little boy.

"There is no class," he said, "whose burdens the phonograph has done so much to lighten as parents. Mothers no longer have to make themselves hoarse telling the children stories on rainy days to keep them out of mischief. It is only necessary to plant the most roguish lad before a phonograph of some nursery classic, to be sure of his whereabouts and his behaviour till the machine runs down, when another set of cylinders can be introduced, and the entertainment carried on. As for the babies, Patti sings mine to sleep at bedtime, and, if they wake up in the night, she is never too drowsy to do it over again. When the children grow too big to be longer tied to their mother's apron-strings, they still remain, thanks to the children's indispensable, though out of her sight, within sound of her voice. Whatever charges or instructions she desires them not to forget, whatever hours or duties she would have them be sure to remember, she depends on the indispensable to remind them of."

At this I cried out. "It is all very well for the mothers," I said, "but the lot of the orphan must seem enviable to a boy compelled to wear about such an instrument of his own subjugation. If boys were what they were in my day, the rate at which their indispensables would get unaccountably lost or broken would be alarming."

Hamage laughed, and admitted that the one he was carrying home was the fourth he had bought for his boy within a month. He agreed with me that it was hard to see how a boy was to get his growth under quite so much government; but his wife, and indeed the ladies generally, insisted that the application of the phonograph to family government was the greatest invention of the age.

Then I asked a question which had repeatedly occurred to me that day,— What had become of the printers?

"Naturally," replied Hamage, "they have had a rather hard time of it. Some classes of books, however, are still printed, and probably will continue to be for some time, although reading, as well as writing, is getting to be an increasingly rare accomplishment."

"Do you mean that your schools do not teach reading and writing?" I exclaimed.

"Oh, yes, they are still taught; but as the pupils need them little after leaving school,—or even in school, for that matter, all their text-books being phonographic,—they usually keep the acquirements about as long as a college graduate does his Greek. There is a strong movement already on foot to drop reading and writing entirely from the school course, but probably a compromise will be made for the present by substituting a shorthand or phonetic system, based upon the direct interpretation of the sound-waves themselves. This is, of course, the only logical method for the visual interpretation of sound. Students and men of research, however, will always need to understand how to read print, as much of the old literature will probably never repay phonographing."

"But," I said, "I notice that you still use printed phrases, as superscriptions, titles, and so forth."

"So we do," replied Hamage, "but phonographic substitutes could be easily devised in these cases, and no doubt will soon have to be supplied in deference to the growing number of those who cannot read."

"Did I understand you," I asked, "that the text-books in your schools even are phonographs?"

"Certainly," replied Hamage; "our children are taught by phonographs, recite to phonographs, and are examined by phonographs."

"Bless my soul!" I ejaculated.

"By all means," replied Hamage; "but there is really nothing to be astonished at. People learn and remember by impressions of sound instead of sight, that is all. The printer is, by the way, not the only artisan whose occupation phonography has destroyed. Since the disuse of print, opticians have mostly gone to the poor-house. The sense of sight was indeed terribly overburdened previous to the introduction of the phonograph, and, now that the sense of hearing is beginning to assume its proper share of work, it would be strange if an improvement in the condition of the people's eyes were not noticeable. Physiologists, moreover, promise us not only an improved vision, but a generally improved physique, especially in respect to bodily carriage, now that reading, writing, and study no longer involves, as formerly, the sedentary attitude with twisted spine and stooping shoulders. The phonograph has at last made it possible to expand the mind without cramping the body."

"It is a striking comment on the revolution wrought by the general introduction of the phonograph," I observed, "that whereas the misfortune of blindness used formerly to be the infirmity which most completely cut a man off from the world of books, which remained open to the deaf, the case is now precisely reversed."

"Yes," said Hamage, "it is certainly a curious reversal, but not so complete as you fancy. By the new improvements in the intensifier, it is expected to enable all, except the stone-deaf, to enjoy the phonograph, even when connected, as on railroad trains, with a common telephonic wire. The stone-deaf will of course be dependent upon printed books prepared for their benefit, as raised-letter books used to be for the blind."

As we entered the elevator to ascend to the upper floors of the establishment, Hamage explained that he wanted me to see, before I left, the process of phonographing books, which was the modern substitute for printing them. Of course, he said, the phonographs of dramatic works were taken at the theatres during the representations of plays, and those of public orations and sermons are either similarly obtained, or, if a revised version is desired, the orator re-delivers his address in the improved form to a phonograph; but the great mass of publications were phonographed by professional elocutionists employed by the large publishing houses, of which this was one. He was acquainted with one of these elocutionists, and was taking me to his room.

We were so fortunate as to find him disengaged. Something, he said, had broken about the machinery, and he was idle while it was being repaired. His work-room was an odd kind of place. It was shaped something like the interior of a rather short egg. His place was on a sort of pulpit in the middle of the small end, while at the opposite end, directly before him, and for some distance along the sides toward the middle, were arranged tiers of phonographs. These were his audience, but by no means all of it. By telephonic communication he was able to address simultaneously other congregations of phonographs in other chambers at any distance. He said that in one instance, where the demand for a popular book was very great, he had charged five thousand phonographs at once with it.

I suggested that the saying of printers, pressmen, bookbinders, and costly machinery, together with the comparative indestructibility of phonographed as compared with printed books, must make them very cheap.

"They would be," said Hamage, "if popular elocutionists, such as Playwell here, did not charge so like fun for their services. The public has taken it into its head that he is the only first-class elocutionist, and won't buy anybody else's work. Consequently the authors stipulate that he shall interpret their productions, and the publishers, between the public and the authors, are at his mercy."

Playwell laughed. "I must make my hay while the sun shines," he said. "Some other elocutionist will be the fashion next year, and then I shall only get hack-work to do. Besides, there is really a great deal more work in my business than people will believe. For example, after I get an author's copy"—.

"Written?" I interjected.

"Sometimes it is written phonetically, but most authors dictate to a phonograph. Well, when I get it, I take it home and study it, perhaps a couple of days, perhaps a couple of weeks, sometimes, if it is really an important work, a month or two, in order to get into sympathy with the ideas, and decide on the proper style of rendering. All this is hard work, and has to be paid for."

At this point our conversation was broken off by Hamage, who declared that, if we were to catch the last train out of town before noon, we had no time to lose.

Of the trip out to Hamage's place I recall nothing. I was, in fact, aroused from a sound nap by the stopping of the train and the bustle of the departing passengers. Hamage had disappeared. As I groped about, gathering up my belongings, and vaguely wondering what had become of my companion, he rushed into the car, and, grasping my hand, gave me an enthusiastic welcome. I opened my mouth to demand what sort of a joke this belated greeting might be intended for, but, on second thought, I concluded not to raise the point. The fact is, when I came to observe that the time was not noon, but late in the evening, and that the train was the one I had left home on, and that I had not even changed my seat in the car since then, it occurred to me that Hamage might not understand allusions to the forenoon we had spent together. Later that same evening, however, the consternation of my host and hostess at my frequent and violent explosions of apparently causeless hilarity left me no choice but to make a clean breast of my preposterous experience. The moral they drew from it was the charming one that, if I would but oftener come to see them, a railroad trip would not so upset my wits.

Edward Bellamy (1850–1898) was an American author who came to prominence through the publication of his utopian science fiction novel "Looking Backward, 2000–1887", published in 1888. The novel was extremely influential at the time, but more for its political ideas than its science fictional elements. Bellamy published a number of SF-related stories, in addition to "With the eyes shut".

Commentary

Brave is the science fiction author who claims to be in the business of prediction. He or she can give us forewarnings, certainly, of possible futures that society should work to avoid, but the field contains few 'futurists' with an enhanced ability to see the shape of things to come. Isaac Asimov, one of the most famous SF writers, liked to point out that he and most of his colleagues wrote about lunar voyages decades before Neil Armstrong set foot on the Moon—it took little foresight to imagine space travel. But no-one in the science fiction community predicted that TV viewers would watch the first Moon landing in the comfort of their living rooms. Asimov, indeed, dismissed his own abilities of prediction: he once wrote a story explaining why no human would ever climb to the summit of Mount Everest. The story appeared some months *after* Hillary and Tenzing reached the top.

Sometimes, though, a brave author speculates in detail about a specific future technology, conceiving of some invention and then listing its ramifications. They write a pure 'prediction' story. Bellamy did this in "With the eyes shut" and his prognostications make for fascinating reading. We sometimes wonder whether our ancestors would adapt to our modern world. Bellamy's example of the 'indispensable' suggests that, if we somehow transported a bunch of Victorians to our time, they would soon adjust.

Today's readers might consider their smartphone an 'indispensable'. Indeed, the term 'indispensable' matches the properties of the smartphone so well that I'll use it here. We of course use an indispensable for making phone calls, a feature Bellamy hinted at. But we also use them to listen to audiobooks, which Bellamy predicted. We use them to watch movies, a feature Bellamy suggested when he mentioned changing pictures of actors. And we use them for a thousand-and-one other things, which Bellamy, writing from his vantage point in 1889, could not appreciate without some training.

Bellamy's story contains imaginative ideas that remain, as far as I know, unrealised in detail but whose kernels have been devised and articulated. For example, he foresees trains playing a soundscape to match the passing landscape. (Back in the 1980s I heard a popular TV historian make this same proposition. Then, I thought it an original proposal. Now, I realise the historian had regurgitated Bellamy's throwaway suggestion.) Well, I have never travelled on a train with that feature. But programmers *have* implemented Bellamy's underlying idea—and improved on it. You can download apps for your indispensable that give you guided audio-tours of historic cities such as London, Paris, or Rome. The apps can work due to an almost magical property of indispensables. The global positioning system, or GPS, can determine your location to within a meter or so. (GPS relies on an understanding of relativity; see Chap. 2.) This allows the apps to know their proximity to Nelson's Column, the Louvre, the Coliseum, or wherever. They can then get your indispensable to play an appropriate commentary.

One aspect Bellamy got altogether wrong resides in the following quote. "Physiologists, moreover, promise us not only an improved vision, but a generally improved physique, especially in respect to bodily carriage, now that reading, writing, and study no longer involves, as formerly, the sedentary attitude with twisted spine and stooping shoulders." Just think of people's posture as they stare into their screens! Think of your own posture!

The main point of Bellamy's story, though, revolved around audiobooks. He speculated that features such as pause, forward, and rewind would help them supplant printed books. Well, audiobooks have indeed become popular. The global market for audiobooks surged during the Covid-19 pandemic, and analysts expect the market volume to have a worth of more than $13 billion by 2030. Even some of my own books have an audio version. (Voice artists, I can confirm, possess bucketloads of talent—much like the story's character Playwell.) And for people with dyslexia or ADHD or low vision; for those who want to 'read' while driving or doing the dishes; for those who wish to enjoy a book together—well, audiobooks are a boon.

So—will Bellamy's prediction prove correct? Will audiobooks replace the printed book?

Forecasters have long predicted the demise of the physical book, but it's worth considering an essay Asimov wrote in 1973. In "The ancient and the ultimate" he pondered the future of videocassette technology and how it might replace the book. He speculated that videocassette devices would become fully mobile. (Your reaction to this might be: "Well, duh!" Half a century ago, though, one needed a plugged-in device to play videocassettes. A mobile consumer device? A science-fiction dream!) The devices of the future would hold a self-contained power source. They would present their images to users without disturbing other people. Controls would become automatic—the device stopping when users looked away and resuming when they looked back. And the devices would allow users to skip around, repeat sections at will, and play at whatever pace they found comfortable. So this was Asimov the futurist describing a perfect device: "self-contained, mobile, non-energy-consuming, perfectly private, and largely under the control of the will". He then pointed out he had described the printed book—a device over five thousand years old! The audiobook perhaps supplements, rather than replaces, the printed book.

What about the well-established ebook market? The digital approach has clear advantages over putting ink on dead trees but I suspect the ebook format, unlike the physical book, will not enjoy a five thousand year history. People will read Shakespeare on paper long after the Kindle has lost favour. Perhaps, though, we will see developments in the form itself. We might see interactive, customizable books with embedded augmented reality; the ability to personalize characters and plots; vastly improved choose-your-own-adventure stories with branching options; and social reading integration. Perhaps we will even see brain–computer interfaces that permit us to experience books in vividly imagined realities.

Perhaps.

But if you forced me to play the futurist I'd say the book—physical sheets of paper, bound together, with a characteristic smell and feel and look—will stay with us for years to come. Books work.

Ongoing technological advances *will* continue to shape one publishing niche, though: scientific publishing.

When I began my career, a scientific article followed a labyrinthine path to publication. First, scientists completed a piece of research. Then they wrote a paper explaining what they had done and how it contributed to our store of knowledge. They (or their secretaries) submitted the paper to a journal. Journal editors with some understanding of the field sent the paper out for peer review. If reviewers accepted the paper—perhaps after demanding the authors make certain changes—copy editors applied the correct journal style and marked up the manuscript for compositors. The compositors re-keyed the manuscript using a dedicated typesetting system and then generated proofs. Journal staff sent the proofs to the authors for checking, and to a different pair of eyes in the publishing house for quality control. Back to the compositor for second proofs. Once all parties had made the necessary corrections, and satisfied themselves of the accuracy of the text, editors collated the paper with others and then published an issue of the journal. Academic libraries subscribed to the journal. After the librarians had collected enough issues of the journal to fill a volume, they would have them bound into a single tome. Those bound volumes then joined others on the groaning shelves of university libraries, resting there, waiting for researchers to open them.

Can you imagine a more expensive, limiting, and time-consuming process?

Today, most science publishers have made the shift from paper to digital. The shift offers advantages. For researchers, it allows for faster publishing. The incorporation of videos and interactive graphs enhances the presentation. The original data can sit alongside the paper, leading to more robust work. For readers, the digital format allows them to access the work in the way they want and to customise it to meet their preferences. They can search by keyword, author, or topic, a feature that can save untold amounts of time and effort. And readers no longer have to make the trip to the university library; they can access the work from anywhere with an internet connection. As with any development, one can identify downsides. In particular, readers of digital science journals might come across fewer opportunities for serendipity, those accidental discoveries made while leafing through a physical journal. But no-one in the scientific community wants to return to the old ways of publishing research.

And what of the future for scientific publishing? Well, publishers will further integrate AI (see Chap. 5) into the process. Artificial intelligence will assist with peer review, identify cases of potential plagiarism, and enhance the discoverability of research. Journals could become platforms for collaboration, where scientists engage in real-time discussions, share data, and work together on projects. And the research described in those journals might have a longer lifespan: since technology allows for post-publication peer review, it permits ongoing evaluation and improvement. How much better than the publishing process of yesterday! Sometimes, once bound, the pages in those dusty volumes in university libraries never again saw daylight.

Notes and Further Reading

story appeared some months after Hillary and Tenzing reached the top—Asimov wrote the story "Everest", in which he suggests the mountain will never be climbed, on 7 April 1953. Hillary and Tenzing, from the British Mount Everest expedition, reached the summit on 29 May. The story did not appear until the December issue of *Universe* magazine, seven months after the successful ascent. Asimov often used this as an example of how poor he was at prediction. See Asimov (1953).

think of people's posture as they stare into their screens—An article by Root (2024) describes how the prevalence of issues relating to smartphone use have led doctors to give the problems nicknames such as 'text neck' and 'text claw'.

global market for audiobooks—Statista (n.d.) has projections on revenue growth in the global audiobook market. Other market research firms have published forecasts that indicate an even greater increase in the popularity and revenue potential of audiobooks.

an essay Asimov wrote in 1973—Asimov argued that reading books had always been a minority pursuit, that it would probably continue to remain so, but that there would always be a sufficient number of readers who preferred physical books for traditional paper publishing to stay viable. See Asimov (1973).

most science publishers have made the shift from paper to digital—The world's first and longest-running scientific journal is *Philosophical Transactions*, launched in March 1665 by Henry Oldenburg, the first secretary of the Royal Society. The journal published Newton's papers on his theory of optics; it published papers by Darwin; it published papers by Hawking. At the time of writing, the latest research to appear in the journal concerns cosmic dust impacts on the Hubble Space Telescope. *Philosophical Transactions* is now fully digital, and the journal's history—from paper to online—is the subject of the book by Fyfe et al. (2022). Scientific publishers across the works have followed a similar journey as that taken by the Royal Society.

References

Asimov, I.: Everest. Universe Science Fiction. December. (1953) Reprinted in *Buy Jupiter and Other Stories*. Doubleday, New York. (1975)

Asimov, I.: The ancient and the ultimate. The Magazine of Fantasy and Science Fiction. January. (1973) Reprinted in The Tragedy of the Moon. Doubleday, New York. (1973)

Fyfe, A., Moxham, N., McDougall-Waters, J., Røstvik, C.M.: A History of Scientific Journals: Publishing at the Royal Society 1665–2015. UCL Press, London (2022)

Root, T.: Do you have text neck? How phones are affecting us physically. The Guardian. 24 January. (2024)

Statista: Audiobooks – worldwide. www.statista.com/outlook/amo/media/books/audiobooks/worldwide (n.d.)

8

Reality: Augmented, Virtual, and Mixed

The Tremendous Adventures of Major Brown (G. K. Chesterton)

Rabelais, or his wild illustrator Gustave Dore, must have had something to do with the designing of the things called flats in England and America. There is something entirely Gargantuan in the idea of economising space by piling houses on top of each other, front doors and all. And in the chaos and complexity of those perpendicular streets anything may dwell or happen, and it is in one of them, I believe, that the inquirer may find the offices of the Club of Queer Trades. It may be thought at the first glance that the name would attract and startle the passer-by, but nothing attracts or startles in these dim immense hives. The passer-by is only looking for his own melancholy destination, the Montenegro Shipping Agency or the London office of the Rutland Sentinel, and passes through the twilight passages as one passes through the twilight corridors of a dream. If the Thugs set up a Strangers' Assassination Company in one of the great buildings in Norfolk Street, and sent in a mild man in spectacles to answer inquiries, no inquiries would be made. And the Club of Queer Trades reigns in a great edifice hidden like a fossil in a mighty cliff of fossils.

    The nature of this society, such as we afterwards discovered it to be, is soon and simply told. It is an eccentric and Bohemian Club, of which the absolute condition of membership lies in this, that the candidate must have invented the method by which he earns his living. It must be an entirely new trade. The exact definition of this requirement is given in the two principal rules. First, it must not be a mere application or variation of an existing trade. Thus, for instance, the Club would not admit an insurance agent simply because instead of insuring men's furniture against being burnt in a fire, he insured, let us say, their trousers against being torn by a mad dog. The principle (as Sir Bradcock Burnaby-Bradcock, in the extraordinarily eloquent and soaring speech to the club on the occasion of the question being raised in the Stormby Smith affair, said wittily and keenly) is the same. Secondly, the trade must be a genuine commercial source of income, the support of its inventor. Thus the Club would not receive a man simply because he chose to pass his days collecting broken sardine tins, unless he could drive a roaring trade in them. Professor Chick made that quite clear. And when one remembers what Professor Chick's own new trade was, one doesn't know whether to laugh or cry.

The discovery of this strange society was a curiously refreshing thing; to realize that there were ten new trades in the world was like looking at the first ship or the first plough. It made a man feel what he should feel, that he was still in the childhood of the world. That I should have come at last upon so singular a body was, I may say without vanity, not altogether singular, for I have a mania for belonging to as many societies as possible: I may be said to collect clubs, and I have accumulated a vast and fantastic variety of specimens ever since, in my audacious youth, I collected the Athenaeum. At some future day, perhaps, I may tell tales of some of the other bodies to which I have belonged. I will recount the doings of the Dead Man's Shoes Society (that superficially immoral, but darkly justifiable communion); I will explain the curious origin of the Cat and Christian, the name of which has been so shamefully misinterpreted; and the world shall know at last why the Institute of Typewriters coalesced with the Red Tulip League. Of the Ten Teacups, of course I dare not say a word. The first of my revelations, at any rate, shall be concerned with the Club of Queer Trades, which, as I have said, was one of this class, one which I was almost bound to come across sooner or later, because of my singular hobby. The wild youth of the metropolis call me facetiously 'The King of Clubs'. They also call me 'The Cherub', in allusion to the roseate and youthful appearance I have presented in my declining years. I only hope the spirits in the better world have as good dinners as I have. But the finding of the Club of Queer Trades has one very curious thing about it. The most curious thing about it is that it was not discovered by me; it was discovered by my friend Basil Grant, a star-gazer, a mystic, and a man who scarcely stirred out of his attic.

Very few people knew anything of Basil; not because he was in the least unsociable, for if a man out of the street had walked into his rooms he would have kept him talking till morning. Few people knew him, because, like all poets, he could do without them; he welcomed a human face as he might welcome a sudden blend of colour in a sunset; but he no more felt the need of going out to parties than he felt the need of altering the sunset clouds. He lived in a queer and comfortable garret in the roofs of Lambeth. He was surrounded by a chaos of things that were in odd contrast to the slums around him; old fantastic books, swords, armour—the whole dust-hole of romanticism. But his face, amid all these quixotic relics, appeared curiously keen and modern—a powerful, legal face. And no one but I knew who he was.

Long ago as it is, everyone remembers the terrible and grotesque scene that occurred in———, when one of the most acute and forcible of the English judges suddenly went mad on the bench. I had my own view of that occurrence; but about the facts themselves there is no question at all. For some months, indeed for some years, people had detected something curious in the judge's conduct. He seemed to have lost interest in the law, in which he had been beyond expression brilliant and terrible as a K.C., and to be occupied in giving personal and moral advice to the people concerned. He talked more like a priest or a doctor, and a very outspoken one at that. The first thrill was probably given when he said to a man who had attempted a crime of passion: "I sentence you to three years imprisonment, under the firm, and solemn, and God-given conviction, that what you require is three months at the seaside." He accused criminals from the bench, not so much of their obvious legal crimes, but of things that had never been heard of in a court of justice, monstrous egoism, lack of humour, and morbidity deliberately encouraged. Things came to a head in that celebrated diamond case in which the Prime Minister himself, that brilliant patrician, had to come forward, gracefully and reluctantly, to give evidence against his valet. After the detailed life of the household had been thoroughly exhibited, the judge requested the Premier again to step forward, which he did with quiet dignity. The judge then said, in a sudden, grating voice: "Get a new soul. That thing's not fit for a dog. Get a new soul." All this, of course, in the eyes of the sagacious, was premonitory of that melancholy and farcical day when his wits actually deserted him in open court. It was a libel case between two very eminent and powerful financiers, against both of whom charges of considerable defalcation were brought. The case was long and complex; the advocates were long and eloquent; but at last, after weeks of work and rhetoric, the time came for the great judge to give a summing-up; and one of his celebrated masterpieces of lucidity and pulverizing logic was eagerly looked for. He had spoken very little during the prolonged affair, and he looked sad and lowering at the end of it. He was silent for a few moments, and then burst into a stentorian song. His remarks (as reported) were as follows:

"O Rowty-owty tiddly-owty Tiddly-owty tiddly-owty Highty-ighty tiddly-ighty Tiddly-ighty ow."

He then retired from public life and took the garret in Lambeth.

I was sitting there one evening, about six o'clock, over a glass of that gorgeous Burgundy which he kept behind a pile of black-letter folios; he was striding about the room, fingering, after a habit of his, one of the great swords in his collection; the red glare of the strong fire struck his square features and his fierce grey hair; his blue eyes were even unusually full of dreams, and he had opened his mouth to speak dreamily, when the door was flung open, and a pale, fiery man, with red hair and a huge furred overcoat, swung himself panting into the room.

"Sorry to bother you, Basil," he gasped. "I took a liberty—made an appointment here with a man—a client—in five minutes—I beg your pardon, sir," and he gave me a bow of apology.

Basil smiled at me. "You didn't know," he said, "that I had a practical brother. This is Rupert Grant, Esquire, who can and does all there is to be done. Just as I was a failure at one thing, he is a success at everything. I remember him as a journalist, a house-agent, a naturalist, an inventor, a publisher, a schoolmaster, a—what are you now, Rupert?"

"I am and have been for some time," said Rupert, with some dignity, "a private detective, and there's my client."

A loud rap at the door had cut him short, and, on permission being given, the door was thrown sharply open and a stout, dapper man walked swiftly into the room, set his silk hat with a clap on the table, and said, "Good evening, gentlemen," with a stress on the last syllable that somehow marked him out as a martinet, military, literary and social. He had a large head streaked with black and grey, and an abrupt black moustache, which gave him a look of fierceness which was contradicted by his sad sea-blue eyes.

Basil immediately said to me, "Let us come into the next room, Gully," and was moving towards the door, but the stranger said:

"Not at all. Friends remain. Assistance possibly."

The moment I heard him speak I remembered who he was, a certain Major Brown I had met years before in Basil's society. I had forgotten altogether the black dandified figure and the large solemn head, but I remembered the peculiar speech, which consisted of only saying about a quarter of each sentence, and that sharply, like the crack of a gun. I do not know, it may have come from giving orders to troops.

Major Brown was a V.C., and an able and distinguished soldier, but he was anything but a warlike person. Like many among the iron men who recovered British India, he was a man with the natural beliefs and tastes of an old maid.

In his dress he was dapper and yet demure; in his habits he was precise to the point of the exact adjustment of a tea-cup. One enthusiasm he had, which was of the nature of a religion—the cultivation of pansies. And when he talked about his collection, his blue eyes glittered like a child's at a new toy, the eyes that had remained untroubled when the troops were roaring victory round Roberts at Candahar.

"Well, Major," said Rupert Grant, with a lordly heartiness, flinging himself into a chair, "what is the matter with you?"

"Yellow pansies. Coal-cellar. P. G. Northover," said the Major, with righteous indignation.

We glanced at each other with inquisitiveness. Basil, who had his eyes shut in his abstracted way, said simply:

"I beg your pardon."

"Fact is. Street, you know, man, pansies. On wall. Death to me. Something. Preposterous."

We shook our heads gently. Bit by bit, and mainly by the seemingly sleepy assistance of Basil Grant, we pieced together the Major's fragmentary, but excited narration. It would be infamous to submit the reader to what we endured; therefore I will tell the story of Major Brown in my own words. But the reader must imagine the scene. The eyes of Basil closed as in a trance, after his habit, and the eyes of Rupert and myself getting rounder and rounder as we listened to one of the most astounding stories in the world, from the lips of the little man in black, sitting bolt upright in his chair and talking like a telegram.

Major Brown was, I have said, a successful soldier, but by no means an enthusiastic one. So far from regretting his retirement on half-pay, it was with delight that he took a small neat villa, very like a doll's house, and devoted the rest of his life to pansies and weak tea. The thought that battles were over when he had once hung up his sword in the little front hall (along with two patent stew-pots and a bad water-colour), and betaken himself instead to wielding the rake in his little sunlit garden, was to him like having come into a harbour in heaven. He was Dutch-like and precise in his taste in gardening, and had, perhaps, some tendency to drill his flowers like soldiers. He was one of those men who are capable of putting four umbrellas in the stand rather than three, so that two may lean one way and two another; he saw life like a pattern in a freehand drawing-book. And assuredly he would not have believed, or even understood, any one who had told him that within a few yards of his brick paradise he was destined to be caught in a whirlpool of incredible adventure, such as he had never seen or dreamed of in the horrible jungle, or the heat of battle.

One certain bright and windy afternoon, the Major, attired in his usual faultless manner, had set out for his usual constitutional. In crossing from one great residential thoroughfare to another, he happened to pass along one of those aimless-looking lanes which lie along the back-garden walls of a row of mansions, and which in their empty and discoloured appearance give one an odd sensation as of being behind the scenes of a theatre. But mean and sulky as the scene might be in the eyes of most of us, it was not altogether so in the Major's, for along the coarse gravel footway was coming a thing which was to him what the passing of a religious procession is to a devout person. A large, heavy man, with fish-blue eyes and a ring of irradiating red beard, was pushing before him a barrow, which was ablaze with incomparable flowers. There were splendid specimens of almost every order, but the Major's own favourite pansies predominated. The Major stopped and fell into conversation, and then into bargaining. He treated the man after the manner of collectors and other mad men, that is to say, he carefully and with a sort of anguish selected the best roots from the less excellent, praised some, disparaged others, made a subtle scale ranging from a thrilling worth and rarity to a degraded insignificance, and then bought them all. The man was just pushing off his barrow when he stopped and came close to the Major.

"I'll tell you what, sir," he said. "If you're interested in them things, you just get on to that wall."

"On the wall!" cried the scandalised Major, whose conventional soul quailed within him at the thought of such fantastic trespass.

"Finest show of yellow pansies in England in that there garden, sir," hissed the tempter. "I'll help you up, sir."

How it happened no one will ever know but that positive enthusiasm of the Major's life triumphed over all its negative traditions, and with an easy leap and swing that showed that he was in no need of physical assistance, he stood on the wall at the end of the strange garden. The second after, the flapping of the frock-coat at his knees made him feel inexpressibly a fool. But the next instant all such trifling sentiments were swallowed up by the most appalling shock of surprise the old soldier had ever felt in all his bold and wandering existence. His eyes fell upon the garden, and there across a large bed in the centre of the lawn was a vast pattern of pansies; they were splendid flowers, but for once it was not their horticultural aspects that Major Brown beheld, for the pansies were arranged in gigantic capital letters so as to form the sentence:

DEATH TO MAJOR BROWN

A kindly looking old man, with white whiskers, was watering them. Brown looked sharply back at the road behind him; the man with the barrow had suddenly vanished. Then he looked again at the lawn with its incredible inscription. Another man might have thought he had gone mad, but Brown did not. When romantic ladies gushed over his V.C. and his military exploits, he sometimes felt himself to be a painfully prosaic person, but by the same token he knew he was incurably sane. Another man, again, might have thought himself a victim of a passing practical joke, but Brown could not easily believe this. He knew from his own quaint learning that the garden arrangement was an elaborate and expensive one; he thought it extravagantly improbable that any one would pour out money like water for a joke against him. Having no explanation whatever to offer, he admitted the fact to himself, like a clear-headed man, and waited as he would have done in the presence of a man with six legs.

At this moment the stout old man with white whiskers looked up, and the watering can fell from his hand, shooting a swirl of water down the gravel path.

"Who on earth are you?" he gasped, trembling violently.

"I am Major Brown," said that individual, who was always cool in the hour of action.

The old man gaped helplessly like some monstrous fish. At last he stammered wildly, "Come down—come down here!"

"At your service," said the Major, and alighted at a bound on the grass beside him, without disarranging his silk hat.

The old man turned his broad back and set off at a sort of waddling run towards the house, followed with swift steps by the Major. His guide led him through the back passages of a gloomy, but gorgeously appointed house, until they reached the door of the front room. Then the old man turned with a face of apoplectic terror dimly showing in the twilight.

"For heaven's sake," he said, "don't mention jackals."

Then he threw open the door, releasing a burst of red lamplight, and ran downstairs with a clatter.

The Major stepped into a rich, glowing room, full of red copper, and peacock and purple hangings, hat in hand. He had the finest manners in the world, and, though mystified, was not in the least embarrassed to see that the only occupant was a lady, sitting by the window, looking out.

"Madam," he said, bowing simply, "I am Major Brown."

"Sit down," said the lady; but she did not turn her head.

She was a graceful, green-clad figure, with fiery red hair and a flavour of Bedford Park. "You have come, I suppose," she said mournfully, "to tax me about the hateful title-deeds."

"I have come, madam," he said, "to know what is the matter. To know why my name is written across your garden. Not amicably either."

He spoke grimly, for the thing had hit him. It is impossible to describe the effect produced on the mind by that quiet and sunny garden scene, the frame for a stunning and brutal personality. The evening air was still, and the grass was golden in the place where the little flowers he studied cried to heaven for his blood.

"You know I must not turn round," said the lady; "every afternoon till the stroke of six I must keep my face turned to the street."

Some queer and unusual inspiration made the prosaic soldier resolute to accept these outrageous riddles without surprise.

"It is almost six," he said; and even as he spoke the barbaric copper clock upon the wall clanged the first stroke of the hour. At the sixth the lady sprang up and turned on the Major one of the queerest and yet most attractive faces he had ever seen in his life; open, and yet tantalising, the face of an elf.

"That makes the third year I have waited," she cried. "This is an anniversary. The waiting almost makes one wish the frightful thing would happen once and for all."

And even as she spoke, a sudden rending cry broke the stillness. From low down on the pavement of the dim street (it was already twilight) a voice cried out with a raucous and merciless distinctness:

"Major Brown, Major Brown, where does the jackal dwell?"

Brown was decisive and silent in action. He strode to the front door and looked out. There was no sign of life in the blue gloaming of the street, where one or two lamps were beginning to light their lemon sparks. On returning, he found the lady in green trembling.

"It is the end," she cried, with shaking lips; "it may be death for both of us. Whenever—".

But even as she spoke her speech was cloven by another hoarse proclamation from the dark street, again horribly articulate.

"Major Brown, Major Brown, how did the jackal die?"

Brown dashed out of the door and down the steps, but again he was frustrated; there was no figure in sight, and the street was far too long and empty for the shouter to have run away. Even the rational Major was a little shaken as he returned in a certain time to the drawing-room. Scarcely had he done so than the terrific voice came:

"Major Brown, Major Brown, where did—".

Brown was in the street almost at a bound, and he was in time—in time to see something which at first glance froze the blood. The cries appeared to come from a decapitated head resting on the pavement.

The next moment the pale Major understood. It was the head of a man thrust through the coal-hole in the street. The next moment, again, it had vanished, and Major Brown turned to the lady. "Where's your coal-cellar?" he said, and stepped out into the passage.

She looked at him with wild grey eyes. "You will not go down," she cried, "alone, into the dark hole, with that beast?"

"Is this the way?" replied Brown, and descended the kitchen stairs three at a time. He flung open the door of a black cavity and stepped in, feeling in his pocket for matches. As his right hand was thus occupied, a pair of great slimy hands came out of the darkness, hands clearly belonging to a man of gigantic stature, and seized him by the back of the head. They forced him down, down in the suffocating darkness, a brutal image of destiny. But the Major's head, though upside down, was perfectly clear and intellectual. He gave quietly under the pressure until he had slid down almost to his hands and knees. Then finding the knees of the invisible monster within a foot of him, he simply put out one of his long, bony, and skilful hands, and gripping the leg by a muscle pulled it off the ground and laid the huge living man, with a crash, along the floor. He strove to rise, but Brown was on top like a cat. They rolled over and over. Big as the man was, he had evidently now no desire but to escape; he made sprawls hither and thither to get past the Major to the door, but that tenacious person had him hard by the coat collar and hung with the other hand to a beam. At length there came a strain in holding back this human bull, a strain under which Brown expected his hand to rend and part from the arm. But something else rent and parted; and the dim fat figure of the giant vanished out of the cellar, leaving the torn coat in the Major's hand; the only fruit of his adventure and the only clue to the mystery. For when he went up and out at the front door, the lady, the rich hangings, and the whole equipment of the house had disappeared. It had only bare boards and white-washed walls.

"The lady was in the conspiracy, of course," said Rupert, nodding. Major Brown turned brick red. "I beg your pardon," he said, "I think not."

Rupert raised his eyebrows and looked at him for a moment, but said nothing. When next he spoke he asked:

"Was there anything in the pockets of the coat?"

"There was sevenpence halfpenny in coppers and a threepenny-bit," said the Major carefully; "there was a cigarette-holder, a piece of string, and this letter," and he laid it on the table. It ran as follows:

Dear Mr. Plover,

I am annoyed to hear that some delay has occurred in the arrangements re Major Brown. Please see that he is attacked as per arrangement tomorrow. The coal-cellar, of course.

Yours faithfully, P. G. Northover.

Rupert Grant was leaning forward listening with hawk-like eyes. He cut in: "Is it dated from anywhere?"

"No—oh, yes!" replied Brown, glancing upon the paper; "14 Tanner's Court, North———"

Rupert sprang up and struck his hands together.

"Then why are we hanging here? Let's get along. Basil, lend me your revolver."

Basil was staring into the embers like a man in a trance; and it was some time before he answered:

"I don't think you'll need it."

"Perhaps not," said Rupert, getting into his fur coat. "One never knows. But going down a dark court to see criminals—"

"Do you think they are criminals?" asked his brother.

Rupert laughed stoutly. "Giving orders to a subordinate to strangle a harmless stranger in a coal-cellar may strike you as a very blameless experiment, but—"

"Do you think they wanted to strangle the Major?" asked Basil, in the same distant and monotonous voice.

"My dear fellow, you've been asleep. Look at the letter."

"I am looking at the letter," said the mad judge calmly; though, as a matter of fact, he was looking at the fire. "I don't think it's the sort of letter one criminal would write to another."

"My dear boy, you are glorious," cried Rupert, turning round, with laughter in his blue bright eyes. "Your methods amaze me. Why, there is the letter. It is written, and it does give orders for a crime. You might as well say that the Nelson Column was not at all the sort of thing that was likely to be set up in Trafalgar Square."

Basil Grant shook all over with a sort of silent laughter, but did not otherwise move.

"That's rather good," he said; "but, of course, logic like that's not what is really wanted. It's a question of spiritual atmosphere. It's not a criminal letter."

"It is. It's a matter of fact," cried the other in an agony of reasonableness.

"Facts," murmured Basil, like one mentioning some strange, far-off animals, "how facts obscure the truth. I may be silly—in fact, I'm off my head—but I never could believe in that man—what's his name, in those capital stories?—Sherlock Holmes. Every detail points to something, certainly; but generally to the wrong thing. Facts point in all directions, it seems to me, like the thousands of twigs on a tree. It's only the life of the tree that has unity and goes up—only the green blood that springs, like a fountain, at the stars."

"But what the deuce else can the letter be but criminal?"

"We have eternity to stretch our legs in," replied the mystic. "It can be an infinity of things. I haven't seen any of them—I've only seen the letter. I look at that, and say it's not criminal."

"Then what's the origin of it?"

"I haven't the vaguest idea."

"Then why don't you accept the ordinary explanation?"

Basil continued for a little to glare at the coals, and seemed collecting his thoughts in a humble and even painful way. Then he said:

"Suppose you went out into the moonlight. Suppose you passed through silent, silvery streets and squares until you came into an open and deserted space, set with a few monuments, and you beheld one dressed as a ballet girl dancing in the argent glimmer. And suppose you looked, and saw it was a man disguised. And suppose you looked again, and saw it was Lord Kitchener. What would you think?"

He paused a moment, and went on:

"You could not adopt the ordinary explanation. The ordinary explanation of putting on singular clothes is that you look nice in them; you would not think that Lord Kitchener dressed up like a ballet girl out of ordinary personal vanity. You would think it much more likely that he inherited a dancing madness from a great grandmother; or had been hypnotised at a seance; or threatened by a secret society with death if he refused the ordeal. With Baden-Powell, say, it might be a bet—but not with Kitchener. I should know all that, because in my public days I knew him quite well. So I know that letter quite well, and criminals quite well. It's not a criminal's letter. It's all atmospheres." And he closed his eyes and passed his hand over his forehead.

Rupert and the Major were regarding him with a mixture of respect and pity. The former said,

"Well, I'm going, anyhow, and shall continue to think—until your spiritual mystery turns up—that a man who sends a note recommending a crime, that is, actually a crime that is actually carried out, at least tentatively, is, in all probability, a little casual in his moral tastes. Can I have that revolver?"

"Certainly," said Basil, getting up. "But I am coming with you." And he flung an old cape or cloak round him, and took a sword-stick from the corner.

"You!" said Rupert, with some surprise, "you scarcely ever leave your hole to look at anything on the face of the earth."

Basil fitted on a formidable old white hat.

"I scarcely ever," he said, with an unconscious and colossal arrogance, "hear of anything on the face of the earth that I do not understand at once, without going to see it."

And he led the way out into the purple night.

We four swung along the flaring Lambeth streets, across Westminster Bridge, and along the Embankment in the direction of that part of Fleet Street which contained Tanner's Court. The erect, black figure of Major Brown, seen from behind, was a quaint contrast to the hound-like stoop and flapping mantle of young Rupert Grant, who adopted, with childlike delight, all the dramatic poses of the detective of fiction. The finest among his many fine qualities was his boyish appetite for the colour and poetry of London. Basil, who walked behind, with his face turned blindly to the stars, had the look of a somnambulist.

Rupert paused at the corner of Tanner's Court, with a quiver of delight at danger, and gripped Basil's revolver in his great-coat pocket.

"Shall we go in now?" he asked.

"Not get police?" asked Major Brown, glancing sharply up and down the street.

"I am not sure," answered Rupert, knitting his brows. "Of course, it's quite clear, the thing's all crooked. But there are three of us, and—"

"I shouldn't get the police," said Basil in a queer voice. Rupert glanced at him and stared hard.

"Basil," he cried, "you're trembling. What's the matter—are you afraid?"

"Cold, perhaps," said the Major, eyeing him. There was no doubt that he was shaking.

At last, after a few moments' scrutiny, Rupert broke into a curse.

"You're laughing," he cried. "I know that confounded, silent, shaky laugh of yours. What the deuce is the amusement, Basil? Here we are, all three of us, within a yard of a den of ruffians—"

"But I shouldn't call the police," said Basil. "We four heroes are quite equal to a host," and he continued to quake with his mysterious mirth.

Rupert turned with impatience and strode swiftly down the court, the rest of us following. When he reached the door of No. 14 he turned abruptly, the revolver glittering in his hand.

"Stand close," he said in the voice of a commander. "The scoundrel may be attempting an escape at this moment. We must fling open the door and rush in."

The four of us cowered instantly under the archway, rigid, except for the old judge and his convulsion of merriment.

"Now," hissed Rupert Grant, turning his pale face and burning eyes suddenly over his shoulder, "when I say 'Four', follow me with a rush. If I say 'Hold him', pin the fellows down, whoever they are. If I say 'Stop', stop. I shall say that if there are more than three. If they attack us I shall empty my revolver on them. Basil, have your sword-stick ready. Now—one, two, three, four!"

With the sound of the word the door burst open, and we fell into the room like an invasion, only to stop dead.

The room, which was an ordinary and neatly appointed office, appeared, at the first glance, to be empty. But on a second and more careful glance, we saw seated behind a very large desk with pigeonholes and drawers of bewildering multiplicity, a small man with a black waxed moustache, and the air of a very average clerk, writing hard. He looked up as we came to a standstill.

"Did you knock?" he asked pleasantly. "I am sorry if I did not hear. What can I do for you?"

There was a doubtful pause, and then, by general consent, the Major himself, the victim of the outrage, stepped forward.

The letter was in his hand, and he looked unusually grim.

"Is your name P. G. Northover?" he asked.

"That is my name," replied the other, smiling.

"I think," said Major Brown, with an increase in the dark glow of his face, "that this letter was written by you." And with a loud clap he struck open the letter on the desk with his clenched fist. The man called Northover looked at it with unaffected interest and merely nodded.

"Well, sir," said the Major, breathing hard, "what about that?"

"What about it, precisely," said the man with the moustache.

"I am Major Brown," said that gentleman sternly.

Northover bowed. "Pleased to meet you, sir. What have you to say to me?"

"Say!" cried the Major, loosing a sudden tempest; "why, I want this confounded thing settled. I want—"

"Certainly, sir," said Northover, jumping up with a slight elevation of the eyebrows. "Will you take a chair for a moment." And he pressed an electric bell just above him, which thrilled and tinkled in a room beyond. The Major put his hand on the back of the chair offered him, but stood chafing and beating the floor with his polished boot.

The next moment an inner glass door was opened, and a fair, weedy, young man, in a frock-coat, entered from within.

"Mr. Hopson," said Northover, "this is Major Brown. Will you please finish that thing for him I gave you this morning and bring it in?"

"Yes, sir," said Mr. Hopson, and vanished like lightning.

"You will excuse me, gentlemen," said the egregious Northover, with his radiant smile, "if I continue to work until Mr. Hopson is ready. I have some books that must be cleared up before I get away on my holiday tomorrow. And we all like a whiff of the country, don't we? Ha! ha!"

The criminal took up his pen with a childlike laugh, and a silence ensued; a placid and busy silence on the part of Mr. P. G. Northover; a raging silence on the part of everybody else.

At length the scratching of Northover's pen in the stillness was mingled with a knock at the door, almost simultaneous with the turning of the handle, and Mr. Hopson came in again with the same silent rapidity, placed a paper before his principal, and disappeared again.

The man at the desk pulled and twisted his spiky moustache for a few moments as he ran his eye up and down the paper presented to him. He took up his pen, with a slight, instantaneous frown, and altered something, muttering—"Careless." Then he read it again with the same impenetrable reflectiveness, and finally handed it to the frantic Brown, whose hand was beating the devil's tattoo on the back of the chair.

"I think you will find that all right, Major," he said briefly.

The Major looked at it; whether he found it all right or not will appear later, but he found it like this:

	£	s.	d.
Major Brown to P. G. Northover			
January 1, to account rendered	5	6	0
May 9, to potting and embedding of 200 pansies	2	0	0
To cost of trolley with flowers	0	15	0
To hiring of man with trolley	0	5	0
To hire of house and garden for one day	1	0	0
To furnishing of room in peacock curtains, copper ornaments, etc.	3	0	0
To salary of Miss Jameson	1	0	0
To salary of Mr. Plover	1	0	0
Total	£14	6	0

A Remittance will oblige.

"What," said Brown, after a dead pause, and with eyes that seemed slowly rising out of his head, "What in heaven's name is this?"

"What is it?" repeated Northover, cocking his eyebrow with amusement. "It's your account, of course."

"My account!" The Major's ideas appeared to be in a vague stampede. "My account! And what have I got to do with it?"

"Well," said Northover, laughing outright, "naturally I prefer you to pay it."

The Major's hand was still resting on the back of the chair as the words came. He scarcely stirred otherwise, but he lifted the chair bodily into the air with one hand and hurled it at Northover's head.

The legs crashed against the desk, so that Northover only got a blow on the elbow as he sprang up with clenched fists, only to be seized by the united rush of the rest of us. The chair had fallen clattering on the empty floor.

"Let me go, you scamps," he shouted. "Let me—"

"Stand still," cried Rupert authoritatively. "Major Brown's action is excusable. The abominable crime you have attempted—"

"A customer has a perfect right," said Northover hotly, "to question an alleged overcharge, but, confound it all, not to throw furniture."

"What, in God's name, do you mean by your customers and overcharges?" shrieked Major Brown, whose keen feminine nature, steady in pain or danger, became almost hysterical in the presence of a long and exasperating mystery. "Who are you? I've never seen you or your insolent tomfool bills. I know one of your cursed brutes tried to choke me—"

"Mad," said Northover, gazing blankly round; "all of them mad. I didn't know they travelled in quartettes."

"Enough of this prevarication," said Rupert; "your crimes are discovered. A policeman is stationed at the corner of the court. Though only a private detective myself, I will take the responsibility of telling you that anything you say—"

"Mad," repeated Northover, with a weary air.

And at this moment, for the first time, there struck in among them the strange, sleepy voice of Basil Grant.

"Major Brown," he said, "may I ask you a question?"

The Major turned his head with an increased bewilderment.

"You?" he cried; "certainly, Mr. Grant."

"Can you tell me," said the mystic, with sunken head and lowering brow, as he traced a pattern in the dust with his sword-stick, "can you tell me what was the name of the man who lived in your house before you?"

The unhappy Major was only faintly more disturbed by this last and futile irrelevancy, and he answered vaguely:

"Yes, I think so; a man named Gurney something—a name with a hyphen—Gurney-Brown; that was it."

"And when did the house change hands?" said Basil, looking up sharply. His strange eyes were burning brilliantly.

"I came in last month," said the Major.

And at the mere word the criminal Northover suddenly fell into his great office chair and shouted with a volleying laughter.

"Oh! it's too perfect—it's too exquisite," he gasped, beating the arms with his fists. He was laughing deafeningly; Basil Grant was laughing voicelessly; and the rest of us only felt that our heads were like weathercocks in a whirlwind.

"Confound it, Basil," said Rupert, stamping. "If you don't want me to go mad and blow your metaphysical brains out, tell me what all this means."

Northover rose.

"Permit me, sir, to explain," he said. "And, first of all, permit me to apologize to you, Major Brown, for a most abominable and unpardonable blunder, which has caused you menace and inconvenience, in which, if you will allow me to say so, you have behaved with astonishing courage and dignity. Of course you need not trouble about the bill. We will stand the loss." And, tearing the paper across, he flung the halves into the waste-paper basket and bowed.

Poor Brown's face was still a picture of distraction. "But I don't even begin to understand," he cried. "What bill? What blunder? What loss?"

Mr. P. G. Northover advanced in the centre of the room, thoughtfully, and with a great deal of unconscious dignity. On closer consideration, there were apparent about him other things beside a screwed moustache, especially a lean, sallow face, hawk-like, and not without a careworn intelligence. Then he looked up abruptly.

"Do you know where you are, Major?" he said.

"God knows I don't," said the warrior, with fervour.

"You are standing," replied Northover, "in the office of the Adventure and Romance Agency, Limited."

"And what's that?" blankly inquired Brown.

The man of business leaned over the back of the chair, and fixed his dark eyes on the other's face.

"Major," said he, "did you ever, as you walked along the empty street upon some idle afternoon, feel the utter hunger for something to happen—something, in the splendid words of Walt Whitman: 'Something pernicious and dread; something far removed from a puny and pious life; something unproved; something in a trance; something loosed from its anchorage, and driving free.' Did you ever feel that?"

"Certainly not," said the Major shortly.

"Then I must explain with more elaboration," said Mr. Northover, with a sigh. "The Adventure and Romance Agency has been started to meet a great modern desire. On every side, in conversation and in literature, we hear of the desire for a larger theatre of events for something to waylay us and lead us splendidly astray. Now the man who feels this desire for a varied life pays a yearly or a quarterly sum to the Adventure and Romance Agency; in return, the Adventure and Romance Agency undertakes to surround him with startling and weird events. As a man is leaving his front door, an excited sweep approaches him and assures him of a plot against his life; he gets into a cab, and is driven to an opium den; he receives a mysterious telegram or a dramatic visit, and is immediately in a vortex of incidents. A very picturesque and moving story is first written by one of the staff of distinguished novelists who are at present hard at work in the adjoining room. Yours, Major Brown (designed by our Mr. Grigsby), I consider peculiarly forcible and pointed; it is almost a pity you did not see the end of it. I need scarcely explain further the monstrous mistake. Your predecessor in your present house, Mr. Gurney-Brown, was a subscriber to our agency, and our foolish clerks, ignoring alike the dignity of the hyphen and the glory of military rank, positively imagined that Major Brown and Mr. Gurney-Brown were the same person. Thus you were suddenly hurled into the middle of another man's story."

"How on earth does the thing work?" asked Rupert Grant, with bright and fascinated eyes.

"We believe that we are doing a noble work," said Northover warmly. "It has continually struck us that there is no element in modern life that is more lamentable than the fact that the modern man has to seek all artistic existence in a sedentary state. If he wishes to float into fairyland, he reads a book; if he wishes to dash into the thick of battle, he reads a book; if he wishes to soar into heaven, he reads a book; if he wishes to slide down the banisters, he reads a book. We give him these visions, but we give him exercise at the same time, the necessity of leaping from wall to wall, of fighting strange gentlemen, of running down long streets from pursuers—all healthy and pleasant exercises. We give him a glimpse of that great morning world of Robin Hood or the Knights Errant, when one great game was played under the splendid sky. We give him back his childhood, that godlike time when we can act stories, be our own heroes, and at the same instant dance and dream."

Basil gazed at him curiously. The most singular psychological discovery had been reserved to the end, for as the little business man ceased speaking he had the blazing eyes of a fanatic.

Major Brown received the explanation with complete simplicity and good humour.

"Of course; awfully dense, sir," he said. "No doubt at all, the scheme excellent. But I don't think—" He paused a moment, and looked dreamily out of the window. "I don't think you will find me in it. Somehow, when one's seen—seen the thing itself, you know—blood and men screaming, one feels about having a little house and a little hobby; in the Bible, you know, 'There remaineth a rest'."

Northover bowed. Then after a pause he said:

"Gentlemen, may I offer you my card. If any of the rest of you desire, at any time, to communicate with me, despite Major Brown's view of the matter—"

"I should be obliged for your card, sir," said the Major, in his abrupt but courteous voice. "Pay for chair."

The agent of Romance and Adventure handed his card, laughing.

It ran, "P. G. Northover, B.A., C.Q.T., Adventure and Romance Agency, 14 Tanner's Court, Fleet Street."

"What on earth is 'C.Q.T.'?" asked Rupert Grant, looking over the Major's shoulder.

"Don't you know?" returned Northover. "Haven't you ever heard of the Club of Queer Trades?"

"There seems to be a confounded lot of funny things we haven't heard of," said the little Major reflectively. "What's this one?"

"The Club of Queer Trades is a society consisting exclusively of people who have invented some new and curious way of making money. I was one of the earliest members."

"You deserve to be," said Basil, taking up his great white hat, with a smile, and speaking for the last time that evening.

When they had passed out the Adventure and Romance agent wore a queer smile, as he trod down the fire and locked up his desk. "A fine chap, that Major; when one hasn't a touch of the poet one stands some chance of being a poem. But to think of such a clockwork little creature of all people getting into the nets of one of Grigsby's tales," and he laughed out aloud in the silence.

Just as the laugh echoed away, there came a sharp knock at the door. An owlish head, with dark moustaches, was thrust in, with deprecating and somewhat absurd inquiry.

"What! back again, Major?" cried Northover in surprise. "What can I do for you?"

The Major shuffled feverishly into the room.

"It's horribly absurd," he said. "Something must have got started in me that I never knew before. But upon my soul I feel the most desperate desire to know the end of it all."

"The end of it all?"

"Yes," said the Major. "'Jackals', and the title-deeds, and 'Death to Major Brown'."

The agent's face grew grave, but his eyes were amused.

"I am terribly sorry, Major," said he, "but what you ask is impossible. I don't know any one I would sooner oblige than you; but the rules of the agency are strict. The Adventures are confidential; you are an outsider; I am not allowed to let you know an inch more than I can help. I do hope you understand—"

"There is no one," said Brown, "who understands discipline better than I do. Thank you very much. Good night."

And the little man withdrew for the last time.

He married Miss Jameson, the lady with the red hair and the green garments. She was an actress, employed (with many others) by the Romance Agency; and her marriage with the prim old veteran caused some stir in her languid and intellectualized set. She always replied very quietly that she had met scores of men who acted splendidly in the charades provided for them by Northover, but that she had only met one man who went down into a coal-cellar when he really thought it contained a murderer.

The Major and she are living as happily as birds, in an absurd villa, and the former has taken to smoking. Otherwise he is unchanged—except, perhaps, there are moments when, alert and full of feminine unselfishness as the Major is by nature, he falls into a trance of abstraction. Then his wife recognizes with a concealed smile, by the blind look in his blue eyes, that he is wondering what were the title-deeds, and why he was not allowed to mention jackals. But, like so many old soldiers, Brown is religious, and believes that he will realize the rest of those purple adventures in a better world.

> **Gilbert Keith Chesterton** (1874–1936) was an English polymath who is perhaps now best known for creating the character of Father Brown, a priest-cum-detective. G.K. Chesterton was one of those authors who wrote on seemingly every imaginable topic in all available formats. Only a fraction of his output was science fictional, but his novel "The Man Who Was Thursday" influenced well known genre writers such as Raymond Lafferty and Gene Wolfe.

Commentary

Chesterton wrote a series of stories involving The Club of Queer Trades. (He used 'queer' in its old sense of 'strange' or 'odd'.) To join the club, a person had to have invented a new way of earning a living. Perhaps today's jobseekers could benefit from adopting the mindset Chesterton had when writing these stories.

When I graduated, more years ago than I care to acknowledge, my teachers expected a large proportion of my class to go on to some sort of job involving physics—research, maybe, or industry, or teaching. But, as responsible professionals, our teachers exposed us to other employment opportunities so that, even if we didn't know how we wanted to earn a living, we would at least understand the contours of the world of work. Several of my peers trained for a career in computing; some became engineers; some got rich, as financial institutions vacuumed up numerate graduates. As it turned out, I spent most of my working life in universities. And, just as my teachers helped me, I have felt a responsibility to help prepare students for employment. The difference between then and now? Those of us in higher education today feel we must prepare our students for jobs that do not yet exist.

Of course, some occupations die out and others appear. That has always happened. One of my distant ancestors made his living as a nail-maker. That noble profession withered when machines started making nails for us. The *pace* at which recent technological developments create new jobs makes today's situation different from previous generations. A dozen years ago, as I write, no-one would have imagined making a living as a social media manager. A dozen months ago, as I write, no-one would have imagined making money from prompt engineering. Today's graduates will move into jobs that none of us yet understand.

In this 1903 story, Chesterton invented a new occupation: 'worldbuilder' for alternate reality games. Then, the job existed in Chesterton's imagination. Now, people earn money from it.

An *alternate* reality game (ARG) takes an interactive, real-world narrative and adapts the details depending on the actions of the players. ARGs have become a mainstream form of entertainment. Consider, for example, the popularity of 'murder mystery' evenings. Players enact a murder, participants try to unmask the killer, and at the end a detective reveals who solved the case. Technology has added new elements to the ARG. Worldbuilders can combine media (TV shows, books, magazines) with puzzles and treasure hunts that take place in the real world. The internet has of course added yet another dimension to ARGs.

A different set of occupations has developed around a similar concept: the *augmented* reality (AR) game. Chesterton would have understood the general idea of AR, but not the details. AR depends on a technology that would seem miraculous to him: the smartphone—or 'indispensable', to use the term we introduced in Chap. 7. An AR game brings elements of game play into the real world. Using the camera on their 'indispensable', players interact with virtual characters or objects in the everyday environment. Pokémon Go

8 Reality: Augmented, Virtual, and Mixed

epitomises the popularity of these games. In Pokémon Go, players engage with virtual creatures in real-world locations. By 2019, 3 years after its release, fans across the globe had downloaded the game more than a billion times.

Workers in the gaming industry pioneered the use of augmented reality, but the ability to overlay digital objects onto the physical world has uses elsewhere. The military has found applications. (One example: AR can help soldiers prepare for high-pressure situations, as illustrated in Fig. 8.1.) Augmented reality also has uses in education. (One example: the camera in an 'indispensable', combined with an AR function, allows students to measure the size of objects and thus explore their environment in new ways.) It has applications in manufacturing. (One example: AR allows a technician to read instructions on a work surface and execute a standardised process step-by-step, thereby saving time, reducing errors, and lessening the cognitive load.) And it has applications in health care. (One example: a handheld AR scanner can project over the skin and show the location of blood vessels, making it easier for a nurse to find a vein on the first attempt.) For each of these sectors,

Fig. 8.1 All arms of the military have found uses for augmented, virtual, and mixed reality. Here, a soldier navigates the 'streets' in the augmented reality laboratory at the Center for Applied Brain and Cognitive Sciences, a joint research initiative between the Soldier Center at CCDC and the School of Engineering at Tufts University. The Center conducts research on individuals and teams working in high-pressure environments. (Credit: David Kamm, CCDC Soldier Center. Public domain)

professionals continue to find new uses for AR. And, as more applications arise, existing jobs change and new ones become possible.

One can extend reality in ways beyond AR. Whereas AR overlays digital information onto the real world, virtual reality (VR) allows a user to experience a fully immersive digital environment. For VR applications, users typically wear a headset. Their environment then becomes visually and audibly distinct from their physical surroundings. Deploy haptic feedback mechanisms too and the environment becomes tangibly different. VR has found uses in gaming; in education and training; in manufacturing; and in healthcare. It even has uses in art galleries. Thanks to VR, I have floated past a three-dimensional representation of one of my favourite paintings, Van Gogh's *Café Terrace at Night*. As with AR, then, the various applications of VR extend the scope of existing jobs and generate new types of job.

At present, a user experiences VR within the confines of a specific context—a particular game or application, say. Some technology companies hope to stitch together these diverse environments. Doing so will permit users to interact in a vast virtual space, the *metaverse*, in way that resembles interaction in the physical world. The idea goes back a long way (at least in computing terms). The term 'metaverse', a portmanteau of 'meta' and 'universe', first appears in the 1992 novel *Snow Crash* by Neal Stephenson. In the novel, the word describes a three-dimensional virtual space where people, via avatars, interact with others and with software agents. In 2003, Linden Lab launched *Second Life*—an online platform that permitted precisely such interaction. (In 2006, I attended a session exploring the potential use of *Second Life* in education. My lack of computer acumen meant my avatar spent two hours stuck underneath a desk—in the real world, two of the most frustrating hours of my life!) *Second Life* remains popular, with about 40,000 active users at any time. Advances in technology have opened further possibilities. In 2021, the company Facebook announced it would focus 'on building the metaverse' and rebranded itself as Meta Platforms.

At the time of this writing, an element of hype surrounds the concept of the metaverse. The term itself risks sliding into meaninglessness, with people using the word in different ways. That lack of clarity perhaps comes from the immaturity of the technology. After all, engineers started constructing the internet in the 1970s. Back then they tried out a variety of approaches, only a selection of which forms part of our current internet. Today's engineers might operate at the same stage with the metaverse, hitting dead ends before finding technological paths that work. Another possibility, of course, sees the metaverse never moving beyond the hype. Even if the concept stagnates, though, advances in VR technology will affect how we access the internet, at least for some of us, some of the time. And that will lead to new jobs. Consider that existing VR walk-through of *Café Terrace at Night*. Perhaps some future 'worldbuilder', a high-tech version of the trade practiced by the character Northover in Chesterton's story, will build on those foundations. Perhaps one day I will stop at the café, sit at one of the tables, talk to the other customers ... and embark on an adventure with one of them.

We have looked at augmented reality (AR) and virtual reality (VR). Mixed reality (MR) blends the two. It combines the digital world with the real world in a way that allows the user to interact with both physical and virtual objects. MR promises to enable engineers to test models in three dimensions before building anything; doctors to manipulate medical scans to help plan operations; students to learn with virtual objects; golfers to improve their swing before trying it on the links—MR has endless applications. That's the hope, anyway. And all these applications, and others not yet imagined, open up the possibility for new jobs.

Perhaps we need to think not of discrete applications such as AR, VR, and MR, but rather of a 'reality continuum'. The alternate reality described by Chesterton has its roots in the physical world, but as we have seen we now have technologies to extend reality (XR): augmented reality overlays digital elements onto the physical world; mixed reality blends the physical with the digital; and virtual reality offers a fully immersive digital environment. Combine alternate reality and extended reality with artificial intelligence and machine learning (discussed in Chap. 5) and we have ... what? Well, in addition to a horrible tangle of letters (XRAIML) we have a playground for science fiction writers—and for those who want to dream up new and wonderful ways of earning a living.

Notes and Further Reading

ARGs have become a mainstream form of entertainment—For more information about the core concepts of alternate reality games see, for example, ARGNet (n.d.).

augmented reality (AR) game—The Interaction Design Foundation–IxDF (n.d.) has more information about augmented reality, and how it differs from virtual and mixed reality.

first appears in the 1992 novel Snow Crash—The novel *Snow Crash* contains numerous fictitious technologies, including a construct the author called the 'metaverse'. See Stephenson (1992).

Linden Lab launched Second Life—To try *Second Life* for yourself, visit Second Life (2024).

think in terms of a 'reality continuum'—After discussing what is meant by mixed reality, augmented reality, and extended reality, Parveau and Adda (2020) propose a classification scheme for XR paradigms.

References

ARGNet: Getting started with ARGs. www.argn.com/getting_started_with_ args/ (n.d.)

Interaction Design Foundation–IxDF. What is augmented reality (AR)? www.interaction-design.org/literature/topics/augmented reality (n.d.)

Parveau, M., Adda, M.: Toward a user-centric classification scheme for extended reality paradigms. J.Ambient Intell. Humanized Comput. **11**(1) (2020). https://doi.org/10.1007/s12652-019-01352-9

Second Life. Homepage. https://secondlife.com (2024)

Stephenson, N.: Snow Crash. Bantam, New York (1992)

9

Pandemic

The Black Grippe (Edgar Wallace)

Dr. Hereford Bevan was looking thoughtfully at a small Cape rabbit; the rabbit took not the slightest notice of Dr. Hereford Bevan. It crouched on a narrow bench, nibbling at a mess of crushed mealies and seemed perfectly content with its lot, in spite of the fact that the bench was situated in the experimental laboratory of the Jackson Institute of Tropical Medicines.

In the young principal's hand was a long porcelain rod with which from time to time he menaced the unconscious feeder, without, however, producing so much as a single shiver of apprehension. With his long ears pricked, his sensitive nostrils quivering—he was used to the man-smell of Hereford Bevan by now—and his big black eyes staring unwinkingly ahead, there was little in the appearance of the rabbit to suggest abnormal condition.

For the third time in a quarter of an hour Bevan raised the rod as though to strike the animal across the nose, and for the third time lowered the rod again. Then with a sigh he lifted the little beast by the ears and carried him, struggling and squirming, to a small hutch, put him in very gently, and closed the wire-netted door.

He stood staring at the tiny inmate and fetched a long sigh. Then he left the laboratory and walked down to the staff study.

Stuart Gold, his assistant, sat at a big desk, pipe in mouth, checking some calculations. He looked up as Bevan came in.

"Well," he said, "what has Bunny done?"

"Bunny is feeding like a pig," said Bevan, irritably.

"No change?"

Bevan shook his head and looked at his watch. "What time—" he began.

"The boat train was in ten minutes ago," said Stuart Gold. "I have been on the 'phone to Waterloo. He may be here at any minute now."

Bevan walked up and down the apartment, his hands thrust into his trousers pockets, his chin on his breast.

Presently he walked to the window and looked out at the busy street. Motor-buses were rumbling past in an endless procession. The sidewalks were crowded with pedestrians, for this was the busiest thoroughfare in the West End of London and it was the hour of the day when the offices were absorbing their slaves.

As he looked, a taxi drew up opposite the door and a man sprang out with all the agility of youth, though the iron-grey whiskers about his chin and the seamed red face placed him amongst the sixties.

"It is he!" cried Hereford Bevan, and dashed from the room to welcome the visitor, taking the portmanteau from his hand.

"It is awfully good of you to come, professor," he said, shaking the traveller warmly by the hand. "Ever since I telegraphed I have been scared sick for fear I brought you on a fool's errand."

"Nonsense," said the elder man, sharply; "I was coming to Europe anyway, and I merely advanced my date of sailing. I'd sooner come by the Mauritania than the slow packet by which I had booked. How are you? You are looking bright."

Hereford Bevan led the newcomer to the study and introduced him to Gold.

Professor Van der Bergh was one of those elderly men who never grow old. His blue eye was as clear as it had been on his twentieth birthday, his sensitive mouth was as ready to smile as ever it had been in the flower of his youth. A professor of pathology, a great anatomist, and one of the foremost bacteriologists in the United States, Bevan's doubts and apprehensions were perhaps justified, though he was relieved in mind to discover that he had merely accelerated the great man's departure from New York and was not wholly responsible for a trip which might end in disappointment.

"Now," said Van der Bergh, spreading his coat-tails and drawing his chair to the little fire, "just give me a second to light my pipe and tell me all your troubles."

He puffed away for a few seconds, blew out the match carefully and threw it into the grate, then before Bevan could speak he said: "I presume that the epidemic of January has scared you?"

Hereford Bevan nodded.

"Well," said the professor, reflectively, "I don't wonder. The 1918 epidemic was bad enough. I am not calling it influenza, because I think very few of us are satisfied to affix that wild label to a devastating disease which appeared in the most mysterious fashion, took its toll, and disappeared as rapidly and mysteriously."

He scratched his beard, staring out of the window.

"I haven't heard any theory about that epidemic which has wholly satisfied me," he said. "People talk glibly of 'carriers' of 'infection', but who infected the wild tribes in the centre of Africa on the very day that whole communities of Eskimos were laid low in parts of the Arctic regions which were absolutely isolated from the rest of the world?"

Bevan shook his head.

"That is the mystery that I have never solved," he said, "and never hope to."

"I wouldn't say that," said the professor, shaking his head. "I am always hoping to get on the track of first causes, however baffling they may be. Anyway, I am not satisfied to describe that outbreak as influenza, and it really does not matter what label we give to it for the moment. You might as truly call it the Plague or the Scourge. Now let's get down to the epidemic of this year. I should like to compare notes with you because I have always found that the reports of this Institute are above suspicion. I suppose it has been suggested to you," he went on, "that the investigation of this particular disease is outside the province of tropical medicines?"

Stuart Gold laughed.

"We are reminded of that every day," he said, dryly.

"Now just tell me what happened in January of this year," said the professor. Dr. Bevan seated himself at the table, pulled open a drawer, and took out a black-covered exercise book.

"I'll tell you briefly," he said, "and without attempting to produce statistics. On the 18th January, as near three o'clock in the afternoon as makes no difference, the second manifestation of this disease attacked this country, and, so far as can be ascertained, the whole of the Continent."

The professor nodded. "What were the symptoms?" he asked.

"People began to cry—that is to say, their eyes filled with water and they felt extremely uncomfortable for about a quarter of an hour. So far as I can discover the crying period did not last much more than a quarter of an hour, in some cases a much shorter time."

Again the professor nodded.

"That is what happened in New York," he said, "and this symptom was followed about six hours later by a slight rise of temperature, shivering, and a desire for sleep."

"Just the same sort of thing happened here," said Bevan, "and in the morning everybody was as well as they had been the previous morning, and the fact that it had occurred might have been overlooked but for the observation made in various hospitals. Gold and I were both stricken at the same time. We both took blood and succeeded in isolating the germs."

The professor jumped up.

"Then you are the only people who have it," he said, "nobody else in the world seems to have taken that precaution."

Stuart Gold lifted a big bell-shaped glass cover from a microscope, took from a locked case a thin microscopic slide, and inserted it in the holder. He adjusted the lens, switched on a shaded light behind the instrument, and beckoned the professor forward.

"Here it is, sir," he said.

Professor Van der Bergh glued his eye to the instrument and looked for a long time.

"Perfect," he said. "I have never seen this fellow before. It looks rather like a trypnasome."

"That's what I told Bevan," said Stuart Gold.

The professor was still looking.

"It is like and it is unlike," he said. "Of course, it is absurd to suggest that you've all had an attack of sleeping sickness, which you undoubtedly would have had if this had been a trypnasome, but surely this bug is a new one to me!"

He walked back to his chair, puffing thoughtfully at his pipe. "What did you do?"

"I made a culture," said Bevan, "and infected six South African rabbits. In an hour they developed the first symptoms. Their eyes watered for the prescribed time, their temperature rose six hours later, and in the morning they were all well."

"Why South African rabbits?" asked Van der Bergh, curiously.

"Because they develop secondary symptoms of any disease at twice the rate of a human being—at least that has been my experience," explained Bevan. "I found it by accident whilst I was in Grahamstown, in South Africa, and it has been a very useful piece of knowledge to me. When I wired to you I had no idea there were going to be any further developments. I merely wanted to make you acquainted with the bug—"

The professor looked up sharply. "Have there been further developments?" he asked, and Bevan nodded.

"Five days ago," he said, speaking slowly, "the second symptom appeared. I will show you." He led the way back to the laboratory, went to the little hutch, and lifted the twisting, struggling rabbit to the bench under a blaze of electric light. The professor felt the animal gingerly.

"He has no temperature," he said, "and looks perfectly normal. What is the matter with him?" Bevan lifted the little beast and held his head toward the light.

"Do you notice anything?" he asked.

"Good heavens!" said Van der Bergh; "he's blind!"

Bevan nodded.

"He's been blind for five days," he said. "But—"

Van der Bergh stared at him. "Do you mean—"

Bevan nodded.

"I mean, that when the secondary symptom comes, and it should come in a fortnight from today."

He stopped.

He had replaced the animal upon the bench and had put out his hand to stroke his ears when suddenly the rabbit groped back from him. Again he reached out his hand and again the animal made a frantic attempt to escape.

"He sees now," said the professor.

"Wait," said Bevan.

He took down a board to which a paper was pinned, looked at his watch, and jotted a note. "Thank God for that," he said; "the blindness lasts for exactly one hundred and twenty hours."

"But do you mean," asked Van der Bergh, with an anxious little frown, "that the whole world is going blind for five days?"

"That is my theory," replied the other.

"Phew!" said the professor, and mopped his face with a large and gaudy handkerchief. They went back without another word to the study and Van der Bergh began his technical test. For his information sheet after sheet of data were placed before him. Records of temperature, of diet and the like were scanned and compared, whilst Bevan made his way to another laboratory to examine the remaining rabbits.

He returned as the professor finished.

"They can all see," he said; "I inspected them this morning and they were as blind as bats."

Presently the professor finished. "I am going down to our Embassy," he said, "and the best thing you boys can do is to see some representative of your Government. Let me see, Sir Douglas Sexton is your big man, isn't he?"

Bevan made a wry face.

"He is the medical gentleman who has the ear of the Government," he said, "but he is rather an impossible person. He's one of the old school—".

"I know that school," said the professor, grimly, "it's a school where you learn nothing and forget nothing. Still, it's your duty to warn him."

Bevan nodded and turned to Stuart Gold.

"Will you cancel my lecture, Gold?" he said; "let Cartwright take the men through that demonstration I gave yesterday. I'll go down and see Sexton though he wither me!"

Sir Douglas Sexton had a large house in a very large square. He was so well-off that he could afford a shabby butler. That shrunken man shook his head when Dr. Bevan made his enquiry.

"I don't think Sir Douglas will see you, sir," he said. "He has a consultation in half an hour's time and he is in his library, with orders that he is not to be disturbed in any circumstances."

"This is a very vital matter and I simply must see Sir Douglas," said Bevan, firmly. The butler was gone for some time and presently returned to usher the caller into a large and gloomy room, where Sir Douglas sat surrounded by open books.

He greeted Bevan with a scowl, for the younger school were not popular with the Sextonians.

"Really, it is most inconvenient, doctor, for you to see me at this moment," he complained, "I suppose you want to ask about the Government grant to the Jackson Institute. I was speaking to the Prime Minister yesterday and he did not seem at all inclined to agree to spend the country's money."

"I haven't come about the grant, Sir Douglas," replied Bevan, "but a matter of much greater importance."

In as few words as possible he gave the result of his experiment, and on the face of Sir Douglas Sexton was undisguised incredulity.

"Come, come," he said, when Dr. Bevan had finished, and permitted his heavy features to relax into a smile. "Now, that sort of stuff is all very well for the Press if you want to make a sensation and advertise your name, but surely you are not coming to me, a medical man, and a medical man, moreover, in the confidence of the Government and the Ministry of Health, with a story of that kind! Of course, there was some sort of epidemic, I admit, on the 18th I myself suffered a little inconvenience, but I think that phenomena could be explained by the sudden change of wind from the southwest to the north-east and the corresponding drop in temperature. You may have noticed that the temperature dropped six degrees that morning."

"I am not bothering about the cause of the epidemic," said Bevan, patiently. "I am merely giving you, Sir Douglas, a rough account of what form the second epidemic will take."

Sir Douglas smiled.

"And do you expect me," he asked with acerbity, "to go to the Prime Minister of England and tell him that in fourteen days the whole of the world is going blind? My dear good man, if you published that sort of story you would scare the people to death and set back the practice of medicine a hundred years! Why, we should all be discredited!"

"Do you think that if I saw the Prime Minister—" began Bevan, and Sir Douglas stiffened.

"If you know the Prime Minister or have any friends who could introduce you," he said, shortly, "I have not the slightest objection to your seeing him. I can only warn you that the Prime Minister is certain to send for me and that I should give an opinion which would be directly contrary to yours. I think you have made a very grave error, Dr. Bevan, and if you were to take the trouble to kill one of your precious rabbits and dissect it you would discover another cause for this blindness."

"The opinion of Dr. Van der Bergh," began Bevan, and Sir Douglas snorted.

"I really cannot allow an American person to teach me my business," he said. "I have nothing to say against American medicines or American surgery, and there are some very charming people in America—I am sure this must be the case. And now, doctor, if you will excuse—"

He turned pointedly to his books and Bevan went out.

For 7 days three men worked most earnestly to enlist the attention of the authorities. They might have given the story to the Press and created a sensation, but neither Bevan nor Van der Bergh favoured this method. Eminent doctors who were consulted took views which were extraordinarily different. Some came to the laboratories to examine the records. Others "pooh-poohed" the whole idea.

"Have you any doubt on the matter yourself?" asked the professor, and Bevan hesitated.

"The only doubt I have, sir," he said, "is whether my calculations as to the time are accurate. I have noticed in previous experiments with these rabbits the disease develops about twice as fast as in the human body, but I am far from satisfied that this rule is invariable."

Van der Bergh nodded.

"My Embassy has wired the particulars to Washington," he said, "and Washington takes a very serious view of your discovery. They are making whatever preparations they can."

He went back to his hotel, promising to call on the morrow. Bevan worked all that day testing the blood of his little subjects, working out tables of reaction, and it was nearly four o'clock when he went to bed.

He slept that night in his room at the Institute. He was a good sleeper, and after winding the clock and drawing down the blind he jumped into bed and in less than five minutes was sound asleep. He awoke with the subconscious feeling that he had slept his usual allowance and was curiously alive and awake. The room was in pitch darkness and he remembered with a frown that he had not gone to bed until four o'clock in the morning. He could not have slept two hours.

He put out his hand and switched on the light to discover the time. Apparently the light was not working.

On his bedside table was a box of matches, his cigarette holder, and his cigarettes. He took the box, struck a light, but nothing happened. He threw away the match and struck another—still nothing happened.

He held the faithless match in his hand and suddenly felt a strange warmth at his fingertips. Then with a cry he dropped the match—it had burnt his fingers!

Slowly he put his legs over the edge of the bed and stood up, groping his way to the window and releasing the spring-blind. The darkness was still complete. He strained his eyes but could not even see the silhouette of the window-frame against the night. Then a church-bell struck the hour … nine, ten, eleven, twelve!

Twelve o'clock! It was impossible that it could be twelve o'clock at night. He gasped. Twelve midday and dark!

He searched for his clothes and began to dress. His window was open, yet from outside came no sound of traffic. London was silent—as silent as the grave.

His window looked out upon the busy thoroughfare in which the Jackson Institute was situated, but there was not so much as the clink of a wheel or the sound of a pedestrian's foot.

He dressed awkwardly, slipping on his boots and lacing them quickly, then groped his way to the door and opened it. A voice outside greeted him. It was the voice of Gold.

"Is that you, Bevan?"

"Yes, it is I, what the dickens—" and then the realization of the catastrophe which had fallen upon the world came to him.

"Blind!" he whispered. "We're all blind!"

Gold had been shell-shocked in the war and was subject to nerve-storms. Presently Bevan heard his voice whimpering hysterically.

"Blind!" he repeated. "What a horrible thing!"

"Steady yourself!" said Bevan, sternly. "It has come! But it's only for five days, Gold. Now don't lose your nerve!"

"Oh, I sha'n't lose my nerve!" said Gold, in a shaky voice. "Only it is rather awful, isn't it? Awful, awful! My God! It's awful!"

"Come down to the study!" said Bevan. "Don't forget the two steps leading down to the landing. There are twenty-four stairs, Gold. Count 'em!"

He was half-way down the stairs when he heard somebody sobbing at the foot and recognized the voice of the old housekeeper who attended to the resident staff. She was whimpering and wailing.

'Shut up!" he said, savagely. "What are you making that infernal row about?"

"Oh, sir," she moaned. "I can't see! I can't see!"

"Nobody can see or will see for five days!" said Bevan. "Keep your nerve, Mrs Moreland." He found his way to the study. He had scarcely reached the room before he heard a thumping on the door which led from the street to the staff quarters. Carefully he manœuvred his way into the hall again, came to the door, and unlocked it.

"Halloa," said a cheery voice outside, "is this the Jackson Institute?"

"Thank God you're safe, professor. You took a risk in coming round."

The professor came in with slow, halting footsteps, and Bevan shut the door behind him. "You know your way, I'll put my hand on your shoulder if you don't mind," said Van der Bergh. "Luckily I took the trouble to remember the route. I've been two hours getting here. Ouch!"

"Are you hurt?" asked Bevan.

"I ran against an infernal motor-bus in the middle of the street. It had been left stranded," said the professor. "I think the blindness is general."

Stuart had stumbled into the room soon after them, had found a chair and sat down upon it.

"Now," said Van der Bergh, briskly, "you've got to find your way to your Government offices and interview somebody in authority. There's going to be hell in the world for the next five days. I hope your calculations are not wrong in that respect, Bevan!"

Hereford Bevan said nothing.

"It is very awkward!" it was Gold's quivering voice that spoke, "but, of course, it'll be all right in a day or two."

"I hope so," said the professor's grim voice. "If it's for five days little harm will be done, but—but if it's for ten days!"

Bevan's heart sank at the doubt in the old man's voice. "If it's for ten days?" he repeated.

"The whole world will be dead," said the professor, solemnly, and there was a deep silence.

"Dead?" whispered Gold and Van der Bergh swung round toward the voice. "What's the matter with you?"

"Shell-shock," muttered Bevan under his breath, and the old man's voice took on a softer note.

"Not all of us, perhaps," he said, "but the least intelligent. Don't you realize what has happened and what will happen? The world is going to starve. We are a blind world, and how shall we find food?"

A thrill of horror crept up Bevan's spine as he realized for the first time just what world-blindness meant.

"All the trains have stopped," the professor went on; "I've been figuring it out in my room this morning just what it means. There are blind men in the signal-boxes and blind men on the engines. All transport has come to a standstill. How are you going to get the food to the people? In a day's time the shops, if the people can reach them, will be sold out and it will be impossible to replenish the local stores. You can neither milk nor reap. All the great power-stations are at a standstill. There is no coal being got out of the mines. Wait, where is your telephone?"

Bevan fumbled for the instrument and passed it in the direction of the professor's voice. A pause, then:

"Take it back," said the professor, "of course, that will not be working. The exchange cannot see!"

Bevan heard a methodical puff-puff and the scent of tobacco came to him, and somehow this brought him comfort. The professor was smoking.

He rose unsteadily to his feet. "Put your hand on my shoulder, professor, and, Gold, take hold of the professor's coat or something."

"Where are you going?" asked Van der Bergh.

"To the kitchen," said Bevan; "there's some food there and I'm starving."

The meal consisted in the main of dry bread, biscuits, and cheese, washed down by water. Then Hereford Bevan began his remarkable pilgrimage.

He left the house, and keeping touch with the railings on his right, reached first Cockspur Street and then Whitehall. Half-way along the latter thoroughfare he thumped into a man and, putting out his hand, felt embossed buttons.

"Halloa," he said, "a policeman?"

"That's right, sir," said a voice; "I've been here since the morning. You're in Whitehall. What has happened, sir? Do you know?"

"It is a temporary blindness which has come upon everybody," said Bevan, speaking quickly. "I am a doctor. Now, constable, you are to tell your friends if you meet them and everybody you do meet that it is only temporary."

"I'm not likely to meet anybody," said the constable. "I've been standing here hardly daring to move since it came."

"What time did it happen?"

"About ten o'clock, as near as I can remember," said the policeman.

"How far from here is Downing Street?"

The constable hesitated.

"I don't know where we are," he said, "but it can't be very far."

Two hours' diligent search, two hours of groping and of stumbling, two hours of discussing with frantic men and women whom he met on the way, brought him to Downing Street.

That journey along Whitehall would remain in his mind a horrible memory for all his days. He heard oaths and sobbings. He heard the wild jabberings of somebody—whether it was man or woman he could not say—who had gone mad under the stress of the calamity, and he came to Downing Street as the clock struck three.

He might have passed the Prime Minister's house but he heard voices and recognized one as that of Sexton.

The great man was moaning his trouble to somebody who spoke in a quiet unemotional voice. "Halloa, Sexton!"

Bevan stumbled toward, and collided with the great physician. "Who is it?" said Sexton.

"It is Hereford Bevan."

"It's the man, Prime Minister, the doctor I spoke to you about." A cool hand took Bevan's.

"Come this way," said the voice; "you had better stay, Sexton, you'll never find your way back."

Bevan found himself led through what he judged to be a large hall and then suddenly his feet struck a heavy carpet.

"I think there's a chair behind you," said the new voice, "sit down and tell me all about it." Dr. Bevan spoke for ten minutes, his host merely interjecting a question here and there.

"It can only last for five days," said the voice, with a quiver of emotion, "and we can only last out that five days. You know, of course, that the food supply has stopped. There is no way of averting this terrible tragedy. Can you make a suggestion?"

"Yes, sir," said Bevan. "there are a number of blind institutes throughout the country. Get in touch with them and let their trained men organize the business of industry. I think it could be done."

There was a pause.

"It might be done," said the voice. "Happily the telegraphs are working satisfactorily, as messages can be taken by sound. The wireless is also working and your suggestion shall be carried out."

The days that followed were days of nightmare, days when men groped and stumbled in an unknown world, shrieking for food. On the evening of the second day the water supply failed. The pumping stations had ceased to work. Happily it rained and people were able to collect water in their mackintosh coats.

Dr. Bevan made several excursions a day and in one of these he met another bold adventurer who told him that part of the Strand was on fire. Somebody had upset a lamp without noticing the fact. The doctor made his way toward the Strand but was forced to turn back by the clouds of pungent smoke which met him.

He and his informant (he was a butcher from Smithfield) locked arms and made their way back to the Institute. By some mischance they took a wrong turning and might have been irretrievably lost but they found a guardian angel in the shape of a woman against whom they blundered.

"The Jackson Institute?" she said. "Oh yes, I can lead you there."

She walked with unfaltering footsteps and with such decision that the doctor thought she had been spared the supreme affliction. He asked her this and she laughed.

"Oh, no," she said, cheerfully. "You see, I've been blind all my life. The Government has put us on point duty at various places to help people who have lost their way."

She told them that, according to her information, big fires were raging in half-a-dozen parts of London. She had heard of no railway collisions and the Prime Minister told her—.

"Told you?" said Bevan in surprise, and again she laughed. "I've met him before, you see," she said. "I am Lord Selbury's daughter, Lillian Selbury." Bevan remembered the name. It is curious that he had pictured her, for all the beauty of her voice, as a sad, middle-aged woman. She took his hand in hers and they walked slowly toward his house.

"You'll think I'm horrid if I say I am enjoying this," she said, "and yet I am. It's so lovely to be able to pity others! Of course, it is very dreadful and it is beginning to frighten me a little, and then there's nobody to tell me how pretty I am, because nobody can see. That is rather a drawback, isn't it?" and she laughed again.

"What does the Government think about this?"

"They are terribly upset," she said, in a graver tone; "you see, they cannot get at the people—they are so used to depending on the newspapers, but there are no newspapers now, and if there were nobody could read them. They have just stopped—You step down from the kerb here and walk twenty-five paces and step up again. We are crossing Whitehall Gardens. They have wonderful faith in this Dr. Bevan."

Hereford Bevan felt himself going red.

"I hope their faith is justified," he said, grimly; "I happen to be the wonderful doctor." He felt her fingers grip him in an uncontrollable spasm of surprise.

"Are you really?" she said, with a new note of interest. "Listen!"

They stopped, and he heard the tinkle of a bell.

"That is one of our people from St Mildreds," she said; "the Government is initiating a system of town-criers. It is the only way we can get news to the people."

Bevan listened and heard the sing-song voice of the crier but could not distinguish what he said. The girl led him to his house and there left him. He felt her hand running down his right arm and wondered why until she took his hand and shook it.

Old Professor Van der Bergh roared a greeting as he came into the room.

"Is that you, Bevan?" he asked. "I've got a knuckle of cold ham here, but be careful how you cut it, otherwise you're going to slice your fingers."

He and Stuart Gold had spent the day feeding the various specimens in the laboratory. The fourth day dawned and in the afternoon came a knock at the door. It was the girl.

"I've been ordered to place myself at your disposal, Dr. Bevan," she said; "the Government may need you."

He spent that day wandering through the deserted streets with the girl at his side and as the hundred and twentieth hour approached he found himself looking forward not so much to the end of the tragic experience which he shared with the world, but to seeing with his own eyes the face of this guide of his. He had slept the clock round and just before ten struck he made his way to the street. He heard Big Ben boom the hour and waited for light, but no light came. Another hour passed and yet another, and his soul was seized with blind panic. Suppose sight never returned, suppose his experiments were altogether wrong and that what happened in the case of the rabbits did not happen to Man! Suppose the blindness was permanent! He groaned at the thought.

The girl was with him, her arm in his, throughout that day. His nerves were breaking, and somehow she sensed this fact and comforted him as a mother might comfort a child. She led him into the park with sure footsteps and walked him up and down, trying to distract his mind from the horror with which it was oppressed.

In the afternoon he was sent for to the Cabinet Council and again told the story of his experiments.

"The hundred and twenty hours are passed, are they not, doctor?" said the Premier's voice.

"Yes, sir," replied Bevan in a low voice, "but it is humanly impossible to be sure that that is the exact time."

No other question was asked him but the terror of his audience came back to him like an aura and shrivelled his very heart.

He did not lie down as was his wont that night, but wandered out alone into the streets of London. It must have been two o'clock in the morning when he came back to find the girl standing on the step talking with Van der Bergh.

She came toward him at the sound of his voice.

"There is another Cabinet meeting, doctor," she said, "will you come with me?"

"I hope I haven't kept you long," he said, brokenly. His voice was husky and so unlike his own that she was startled.

"You're not to take this to heart, Dr. Bevan," she said, severely, as they began their pilgrimage to Whitehall. "There's a terrible task waiting for the world which has to be faced."

"Wait, wait!" he said, hoarsely, and gripped the rail with one hand and her arm with the other. Was it imagination? It was still dark, a fine drizzle of rain was falling, but the blackness was dappled with tones of less blackness. There was a dark, straight thing before him, something that seemed to hang in the centre of his eye, and a purple shape beyond, and he knew that he was looking at a London street, at a London lamppost, with eyes that saw. Black London, London devoid of light, London whose streets were packed with motionless vehicles that stood just where they had stopped on the day the darkness fell, London with groping figures half mad with joy, shrieking and sobbing their relief—he drew a long breath.

"What is it? What is it?" said the girl in a frightened voice.

"I can see! I can see!" said Bevan in a whisper.

"Can you?" she said, wistfully. "I—I am so glad. And now—"

He was near to tears and his arms went about her. He fumbled in his pocket for a match and struck a light. That blessed light he saw, and saw, too, the pale spiritual face turned up to his.

"I can see you," he whispered again. "My God! You're the most beautiful thing I have ever seen!"

London slept from sheer force of habit and woke with the grey dawn to see—to look out upon a world that had been lost for five and a half days, but in the night all the forces of the law and the Crown had been working at feverish pace, railways had dragged their drivers from their beds, carriers and stokers had been collected by the police, and slowly the wheels of life were turning again, and a humble world, grateful for the restoration of its greatest gift, hungered in patience and was happy.

> **Edgar Wallace** (1875–1932), an English author, wrote science fiction but is much better known as a prolific writer of thrillers and adventure stories. He published over 170 novels and almost one thousand pieces of short fiction. His writing technique was unusual for the time: he narrated his stories onto wax cylinders and had a secretary type up the text. Wallace could dictate a novel in a few days. He seldom bothered to edit his work.

Commentary

In 2019 I published an anthology similar to this one, and in it I included Jack London's story "The scarlet plague" in a chapter about pandemic. Less than six months after the book appeared the world entered its first major pandemic in a century. Impeccable timing on my part, but I wish it had been different. As of April 2024, the global death toll from Covid-19 stood at more than seven million. Since we have ourselves experienced the scourge of pandemic disease, it seems appropriate to once more include a pandemic story.

Wallace's "The black grippe" appeared in *The Strand* magazine in March 1920, two years after an influenza epidemic began that killed between 17–100 million people. The disease devastated communities across the globe. Wallace (or at least one of Wallace's characters) remained unsure of the cause of such a tragic death toll. In the story, Professor Van der Bergh states: "The 1918 epidemic was bad enough. I am not calling it influenza, because I think very few of us are satisfied to affix that wild label to a devastating disease which appeared in the most mysterious fashion, took its toll, and disappeared as rapidly and mysteriously." The aftermath of the 1918 epidemic, it seems, led some to become 'influenza deniers'.

One can forgive the people of a century ago for feeling uncertain about a disease that had caused so much havoc. The science of microbiology had not yet provided the insights into disease causation that we now take for granted. Medicine, less sophisticated back then, offered little in the way of efficacious treatment. News travelled at a slower pace. No surprise, then, that people did not know what had hit them. On top of all that, the disease struck after the horrors of the First World War. Perhaps those affected by the epidemic did not dwell upon it by design. In this case ignorance provided not bliss, but rather a coping mechanism. One can understand the existence of 'influenza deniers'.

A century later, we have 'Covid deniers'. How can we explain that fact? After all, a century of medical and technological development meant we had better tools this time round.

In December 2019, Li Wenliang, a Chinese eye doctor, warned about a pneumonia-like illness striking people in Wuhan. On 10 January 2020, the World Health Organisation began talking about a "novel coronavirus". On 20 January, scientists confirmed the first case of the coronavirus in America. On 23 January, authorities placed Wuhan—a city of 11 million citizens—under lockdown. People around the world watched this happening: the reach of our communication systems enabled us to follow the spread of the disease in real time. We learned how microbiologists identified the causative agent, SARS-Cov-2 (see Fig. 9.1). And we (well, most of us) marvelled as scientists began to develop safe, effective vaccines.

The timeline of all this should not generate any controversy. And yet most of us will know someone who questions the safety and effectiveness of the

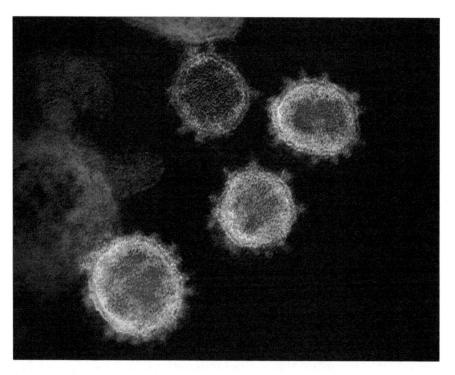

Fig 9.1 A colourised transmission electron microscope image of SARS-Cov-2, the virus that causes the disease Covid-19. The image shows the so-called spike protein protruding from the virus like spikes from a crown. The Latin word for 'crown' is 'corona', hence the term 'coronavirus'. The image was taken at NIAID's Rocky Mountain Laboratories in Montana. (Credit: NIAID, CC BY-SA 2.0 Generic)

vaccines. Some people dispute the scale of Covid-19. A small group even doubts the existence of the disease. One can of course question the policy decisions taken in response to the pandemic, and one can ask whether the undoubted economic and social costs of lockdowns outweighed their benefits. But why should people doubt the evidence of science, news reports and, indeed, their own eyes?

Answering those questions would take behavioural scientists and those interested in the science of persuasion (see Chap. 5) an entire book. Or perhaps several books. One relevant factor, though, surfaces in Wallace's story: failure to communicate uncertainty.

Uncertainty constitutes the natural state of science. After all, that's why scientists do research: they want to find something out. And after they have completed a piece of research, scientists must estimate how much confidence we should all have in their findings. Long years of study and training mean scientists understand that science does not trade in certainties. Indeed, the way science deals with uncertainty represents a key strength. In public discussion, though, those unfamiliar with science often attack that uncertainty. They wrongly equate uncertainty with unreliability. Most people crave certainty, and if they don't get it they can respond in negative ways. Some people worry. Some become cynical about science. And some turn to people who claim certainty—bad actors who spread misinformation and disinformation online.

Humanity faces several existential challenges that science can help address, including human-induced climate change, developments in artificial intelligence, and, of course, pandemics—whether plague, influenza, or some new disease nature throws at us. When discussing these challenges, professionals must convey—with clarity and cogency—information involving scientific uncertainty. A difficult task. But vital.

In *The Uncertainty Handbook*, Corner and co-authors share a dozen tips about communicating uncertainty. Although climate scientists form the primary audience for the *Handbook* (Corner, indeed, lead on the first communications handbook for IPCC authors) those involved in communicating scientific messages around pandemics will also find the tips useful. Try using these five key approaches when you next find yourself in discussion with a Covid denier.

First, emphasise that science is an ongoing debate rather than a set of answers.

People's thinking about science goes awry when they equate science with a set of facts. Science gives us a method for asking questions about the world. As a useful byproduct of the method we gain knowledge. Scientists do not know *everything* about Covid; it does not follow that they know *nothing*.

When people comprehend that fact, and they begin to appreciate the scientific method, they tend not to dismiss out-of-hand messages containing uncertainty.

Second, when talking to the public, start with what you know rather than what you don't know.

Too often, scientists provide caveats before presenting the take-home message—an appropriate approach when you give a seminar to peers but not when trying to communicate, in a limited time and often under a constrained format, with the general public. You do not sacrifice honesty when you speak with clarity. On some matters the science is effectively settled. You should state that. Clearly.

Third, be clear about the scientific consensus.

More than 97% of climate scientists agree that human activity causes climate change, but most people do not know of that consensus. (Mainstream media often adopt a wrong-headed 'both-sides' approach. Small wonder the public might not appreciate the situation.) If you represent that 97% figure on a pie chart then people can see how much agreement there is among scientists. The same goes for health-related information in a pandemic: drive home the consensus!

Fourth, talk about 'risk' rather than 'uncertainty'.

People deal with risk in their everyday lives. Common examples of risk management provide a useful means of presenting comparisons.

Fifth, if you do talk of uncertainty, frame it in a positive rather than a negative way.

A positive framing creates hope. A negative framing creates a feeling of hopelessness. Corner and his co-authors provide the following example from climate change discourse.

- "If we act now, the chance of destructive winter floods occurring is 20%." (Positive frame)
- "If we fail to act, the chance of destructive winter floods occurring is 80%." (Negative frame)

The two sentences carry the same idea, but if you want to induce action rather than despair then choose the positive frame. During a pandemic, one might want citizens to avoid public areas, wear a mask, and isolate if symptomatic. A positive framing of that message:

- "If you practice these three simple steps you can protect yourself and others."

A negative framing of the message:

- "If you do not practice these three steps you can put yourself and others in danger."

If nothing else, the positive frame causes less anxiety.

The Uncertainty Handbook packs these and similar tips into just 14 pages of text. I recommend it.

In science, the task of communicating uncertainty to the public during a pandemic requires skills few scientists have developed. As a community, though, we should learn how best to take on the task because little uncertainty surrounds the following prediction: another pandemic will strike. We don't know what form it will take nor when it will come. But it *will* happen.

Notes and Further Reading

an anthology similar to this one—My book *New Light Through Old Windows* (Webb 2019) contained twelve stories by authors such as Arthur Conan Doyle, Jack London, and H. G. Wells. For the current volume I decided to include stories from different writers and, for the most part, I wanted to cover different areas of science and technology. The events of 2020–21, however, caused me to revisit the chapter on pandemics.

the global death toll—The Worldometer Coronavirus Tracker delivered accurate and timely global statistics on Covid-19 to users and institutions worldwide. The site stopped updating the Tracker on 13 April 2024 because most countries had stopped reporting. Historical data remain accessible, however; see Worldometer (n.d.).

two years after an influenza epidemic—The 'Spanish flu' is the subject of numerous books. See for example Spinney (2017).

we have 'Covid deniers'—Since the Covid-19 pandemic, a great deal of research has gone into understanding the phenomenon of pandemic denial. See for example Thagard (2021) and Rothmund et al. (2022).

Li Wenliang, a Chinese eye doctor—The Centers for Disease Control and Prevention have an online timeline of the Covid-19 pandemic. See CDC (n.d.). Mentioned on the timeline is Li Wenliang (1986–2020). Li Wenliang was an ophthalmologist, who drew attention to cases of people presenting with an unexplained pneumonia-like disease in the week commencing 23 December 2019. The authorities condemned him as a whistleblower. Li Wenliang contracted Covid-19 in January 2020. His death was announced on 7 February. In April, the authorities honoured him as a 'martyr', the highest civilian honour for a person who dies in the service of China.

a dozen tips about communicating certainty—For details of the *Handbook*, see Corner et al. (2015). The IPCC communications handbook mentioned in the text is by Corner et al. (2018). For a flavour of the research that takes place into the problems involved in communicating scientific uncertainty, see for example Doyle et al. (2023).

References

CDC: CDC Museum COVID-19 Timeline. www.cdc.gov/museum/timeline/covid19.html (n.d.)

Corner, A., Lewandowsky, S., Phillips, M., Roberts, O.: The Uncertainty Handbook. University of Bristol, Bristol (2015)

Corner, A., Shaw, C., Clarke, J.: Principles for Effective Communication and Public Engagement on Climate Change: A Handbook for IPCC Authors. Climate Outreach, Oxford (2018)

Doyle, E.E.H., Thompson, J., Hill, S., Williams, M., Paton, D., Harrison, S., Bostrom, A., Becker, J.: Where does scientific uncertainty come from, and from whom? Mapping perspectives of natural hazards science advice. Int. J. Disaster Risk Reduction. **96**, 103948 (2023)

Rothmund, T., Farkhari, F., Ziemer, C.-T., Azevedo, F.: Psychological underpinnings of pandemic denial—patterns of disagreement with scientific experts in the German public during the COVID-19 pandemic. Public Underst. Sci. **31**(4), 437–457 (2022). https://doi.org/10.1177/09636625211068131

Spinney, L.: Pale Rider: The Spanish Flu of 1918 and How It Changed the World. Public Affairs, New York (2017)

Thagard, P.: The cognitive science of COVID-19: Acceptance, denial, and belief change. Methods. **195**, 92–102 (2021). https://doi.org/10.1016/j.ymeth.2021.03.009

Webb, S.: New Light Through Old Windows: Exploring Contemporary Science Through 12 Classic Science Fiction Tales. Springer, Berlin (2019)

Worldometer (n.d.) Coronavirus death toll. www.worldometers.info/coronavirus/coronavirus-death-toll/

10

Suspended Animation

Pausodyne: A Great Chemical Discovery (J. Arbuthnot Wilson)

Walking along the Strand one evening last year towards Pall Mall, I was accosted near Charing Cross Station by a strange-looking, middle-aged man in a poor suit of clothes, who surprised and startled me by asking if I could tell him from what inn the coach usually started for York.

"Dear me!" I said, a little puzzled. "I didn't know there was a coach to York. Indeed, I'm almost certain there isn't one."

The man looked puzzled and surprised in turn. "No coach to York?" he muttered to himself, half inarticulately. "No coach to York? How things have changed! I wonder whether nobody ever goes to York nowadays!"

"Pardon me," I said, anxious to discover what could be his meaning; "many people go to York every day, but of course they go by rail."

"Ah, yes," he answered softly, "I see. Yes, of course, they go by rail. They go by rail, no doubt. How very stupid of me!" And he turned on his heel as if to get away from me as quickly as possible.

I can't exactly say why, but I felt instinctively that this curious stranger was trying to conceal from me his ignorance of what a railway really was. I was quite certain from the way in which he spoke that he had not the slightest conception what I meant, and that he was doing his best to hide his confusion by pretending to understand me. Here was indeed a strange mystery. In the latter end of this nineteenth century, in the metropolis of industrial England, within a stone's-throw of Charing Cross terminus, I had met an adult Englishman who apparently did not know of the existence of railways. My curiosity was too much piqued to let the matter rest there. I must find out what he meant by it. I walked after him hastily, as he tried to disappear among the crowd, and laid my hand upon his shoulder, to his evident chagrin.

"Excuse me," I said, drawing him aside down the corner of Craven Street; "you did not understand what I meant when I said people went to York by rail?"

He looked in my face steadily, and then, instead of replying to my remark, he said slowly, "Your name is Spottiswood, I believe?"

Again I gave a start of surprise. "It is," I answered; "but I never remember to have seen you before."

"No," he replied dreamily; "no, we have never met till now, no doubt; but I knew your father, I'm sure; or perhaps it may have been your grandfather."

"Not my grandfather, certainly," said I, "for he was killed at Waterloo."

"At Waterloo! Indeed! How long since, pray?"

I could not refrain from laughing outright. "Why, of course," I answered, "in 1815. There has been nothing particular to kill off any large number of Englishmen at Waterloo since the year of the battle, I suppose."

"True," he muttered, "quite true; so I should have fancied." But I saw again from the cloud of doubt and bewilderment which came over his intelligent face that the name of Waterloo conveyed no idea whatsoever to his mind.

Never in my life had I felt so utterly confused and astonished. In spite of his poor dress, I could easily see from the clear-cut face and the refined accent of my strange acquaintance that he was an educated gentleman—a man accustomed to mix in cultivated society. Yet he clearly knew nothing whatsoever about railways, and was ignorant of the most salient facts in English history. Had I suddenly come across some Caspar Hauser, immured for years in a private prison, and just let loose upon the world by his gaolers? Or was my mysterious stranger one of the Seven Sleepers of Ephesus, turned out unexpectedly in modern costume on the streets of London? I don't suppose there exists on earth a man more utterly free than I am from any tinge of superstition, any lingering touch of a love for the miraculous; but I confess for a moment I felt half inclined to suppose that the man before me must have drunk the elixir of life, or must have dropped suddenly upon earth from some distant planet.

The impulse to fathom this mystery was irresistible. I drew my arm through his. "If you knew my father," I said, "you will not object to come into my chambers and take a glass of wine with me."

"Thank you," he answered half suspiciously; "thank you very much. I think you look like a man who can be trusted, and I will go with you."

We walked along the Embankment to Adelphi Terrace, where I took him up to my rooms, and seated him in my easy-chair near the window. As he sat down, one of the trains on the Metropolitan line whirred past the Terrace, snorting steam and whistling shrilly, after the fashion of Metropolitan engines generally. My mysterious stranger jumped back in alarm, and seemed to be afraid of some immediate catastrophe. There was absolutely no possibility of doubting it. The man had obviously never seen a locomotive before.

"Evidently," I said, "you do not know London. I suppose you are a colonist from some remote district, perhaps an Australian from the interior somewhere, just landed at the Tower?"

"No, not an Austrian"—I noted his misapprehension—"but a Londoner born and bred."

"How is it, then, that you seem never to have seen an engine before?"

"Can I trust you?" he asked in a piteously plaintive, half-terrified tone. "If I tell you all about it, will you at least not aid in persecuting and imprisoning me?"

I was touched by his evident grief and terror. "No," I answered, "you may trust me implicitly. I feel sure there is something in your history which entitles you to sympathy and protection."

"Well," he replied, grasping my hand warmly, "I will tell you all my story; but you must be prepared for something almost too startling to be credible."

"My name is Jonathan Spottiswood," he began calmly.

Again I experienced a marvellous start: Jonathan Spottiswood was the name of my great-great-uncle, whose unaccountable disappearance from London just a century since had involved our family in so much protracted litigation as to the succession to his property. In fact, it was Jonathan Spottiswood's money which at that moment formed the bulk of my little fortune. But I would not interrupt him, so great was my anxiety to hear the story of his life.

"I was born in London," he went on, "in 1750. If you can hear me say that and yet believe that possibly I am not a madman, I will tell you the rest of my tale; if not, I shall go at once and for ever."

"I suspend judgment for the present," I answered. "What you say is extraordinary, but not more extraordinary perhaps than the clear anachronism of your ignorance about locomotives in the midst of the present century."

"So be it, then. Well, I will tell you the facts briefly in as few words as I can. I was always much given to experimental philosophy, and I spent most of my time in the little laboratory which I had built for myself behind my father's house in the Strand. I had a small independent fortune of my own, left me by an uncle who had made successful ventures in the China trade; and as I was indisposed to follow my father's profession of solicitor, I gave myself up almost entirely to the pursuit of natural philosophy, following the researches of the great Mr. Cavendish, our chief English thinker in this kind, as well as of Monsieur Lavoisier, the ingenious French chemist, and of my friend Dr. Priestley, the Birmingham philosopher, whose new theory of phlogiston I have been much concerned to consider and to promulgate. But the especial subject to which I devoted myself was the elucidation of the nature of fixed air. I do not know how far you yourself may happen to have heard respecting these late discoveries in chemical science, but I dare venture to say that you are at least acquainted with the nature of the body to which I refer."

"Perfectly," I answered with a smile, "though your terminology is now a little out of date. Fixed air was, I believe, the old-fashioned name for carbonic acid gas."

"Ah," he cried vehemently, "that accursed word again! Carbonic acid has undone me, clearly. Yes, if you will have it so, that seems to be what they call it in this extraordinary century; but fixed air was the name we used to give it in our time, and fixed air is what I must call it, of course, in telling you my story. Well, I was deeply interested in this curious question, and also in some of the results which I obtained from working with fixed air in combination with a substance I had produced from the essential oil of a weed known to us in England as lady's mantle, but which the learned Mr. Carl Linnæus describes in his system as Alchemilla vulgaris. From that weed I obtained an oil which I combined with a certain decoction of fixed air into a remarkable compound; and to this compound, from its singular properties, I proposed to give the name of Pausodyne. For some years I was almost wholly engaged in investigating the conduct of this remarkable agent; and lest I should weary you by entering into too much detail, I may as well say at once that it possessed the singular power of entirely suspending animation in men or animals for several hours together. It is a highly volatile oil, like ammonia in smell, but much thicker in gravity; and when held to the nose of an animal, it causes immediate stoppage of the heart's action, making the body seem quite dead for long periods at a time. But the moment a mixture of the Pausodyne with oil of vitriol and gum resin is presented to the nostrils, the animal instantaneously revives exactly as before, showing no evil effects whatsoever from its temporary simulation of death. To the reviving mixture I have given the appropriate name of Anegeiric.

"Of course you will instantly see the valuable medical applications which may be made of such an agent. I used it at first for experimenting upon the amputation of limbs and other surgical operations. It succeeded admirably. I found that a dog under the influence of Pausodyne suffered his leg, which had been broken in a street accident, to be set and spliced without the slightest symptom of feeling or discomfort. A cat, shot with a pistol by a cruel boy, had the bullet extracted without moving a muscle. My assistant, having allowed his little finger to mortify from neglect of a burn, permitted me to try the effect of my discovery upon himself; and I removed the injured joints while he remained in a state of complete insensibility, so that he could hardly believe afterwards in the actual truth of their removal. I felt certain that I had invented a medical process of the very highest and greatest utility.

"All this took place in or before the year 1781. How long ago that may be according to your modern reckoning I cannot say; but to me it seems hardly more than a few months since. Perhaps you would not mind telling me the date of the current year. I have never been able to ascertain it."

"This is 1881," I said, growing every moment more interested in his tale.

"Thank you. I gathered that we must now be somewhere near the close of the nineteenth century, though I could not learn the exact date with certainty. Well, I should tell you, my dear sir, that I had contracted an engagement about the year 1779 with a young lady of most remarkable beauty and attractive mental gifts, a Miss Amelia Spragg, daughter of the well-known General Sir Thomas Spragg, with whose achievements you are doubtless familiar. Pardon me, my friend of another age, pardon me, I beg of you, if I cannot allude to this subject without emotion after a lapse of time which to you doubtless seems like a century, but is to me a matter of some few months only at the utmost. I feel towards her as towards one whom I have but recently lost, though I now find that she has been dead for more than eighty years." As he spoke, the tears came into his eyes profusely; and I could see that under the external calmness and quaintness of his eighteenth century language and demeanour his whole nature was profoundly stirred at the thought of his lost love.

"Look here," he continued, taking from his breast a large, old-fashioned gold locket containing a miniature; "that is her portrait, by Mr. Walker, and a very truthful likeness indeed. They left me that when they took away my clothes at the Asylum, for I would not consent to part with it, and the physician in attendance observed that to deprive me of it might only increase the frequency and violence of my paroxysms. For I will not conceal from you the fact that I have just escaped from a pauper lunatic establishment."

I took the miniature which he handed me, and looked at it closely. It was the picture of a young and beautiful girl, with the features and costume of a Sir Joshua. I recognized the face at once as that of a lady whose portrait by Gainsborough hangs on the walls of my uncle's dining-room at Whittingham

Abbey. It was strange indeed to hear a living man speak of himself as the former lover of this, to me, historic personage.

"Sir Thomas, however," he went on, "was much opposed to our union, on the ground of some real or fancied social disparity in our positions; but I at last obtained his conditional consent, if only I could succeed in obtaining the Fellowship of the Royal Society, which might, he thought, be accepted as a passport into that fashionable circle of which he was a member. Spurred on by this ambition, and by the encouragement of my Amelia, I worked day and night at the perfectioning of my great discovery, which I was assured would bring not only honour and dignity to myself, but also the alleviation and assuagement of pain to countless thousands of my fellow-creatures. I concealed the nature of my experiments, however, lest any rival investigator should enter the field with me prematurely, and share the credit to which I alone was really entitled. For some months I was successful in my efforts at concealment; but in March of this year—I mistake; of the year 1781, I should say—an unfortunate circumstance caused me to take special and exceptional precautions against intrusion.

"I was then conducting my experiments upon living animals, and especially upon the extirpation of certain painful internal diseases to which they are subject. I had a number of suffering cats in my laboratory, which I had treated with Pausodyne, and stretched out on boards for the purpose of removing the tumours with which they were afflicted. I had no doubt that in this manner, while directly benefiting the animal creation, I should indirectly obtain the necessary skill to operate successfully upon human beings in similar circumstances. Already I had completely cured several cats without any pain whatsoever, and I was anxious to proceed to the human subject. Walking one morning in the Strand, I found a beggar woman outside a gin-shop, quite drunk, with a small, ill-clad child by her side, suffering the most excruciating torments from a perfectly remediable cause. I induced the mother to accompany me to my laboratory, and there I treated the poor little creature with Pausodyne, and began to operate upon her with perfect confidence of success.

"Unhappily, my laboratory had excited the suspicion of many ill-disposed persons among the low mob of the neighbourhood. It was whispered abroad that I was what they called a vivisectionist; and these people, who would willingly have attended a bull-baiting or a prize fight, found themselves of a sudden wondrous humane when scientific procedure was under consideration. Besides, I had made myself unpopular by receiving visits from my friend Dr. Priestley, whose religious opinions were not satisfactory to the strict orthodoxy of St. Giles's. I was rumoured to be a philosopher, a torturer of live animals, and an atheist. Whether the former accusation were true or not, let others decide; the two latter, heaven be my witness, were wholly unfounded. However, when the neighbouring rabble saw a drunken woman with a little girl entering my door, a report got abroad at once that I was going to vivisect a Christian child. The mob soon collected in force, and broke into the laboratory. At that moment I was engaged, with my assistant, in operating upon the girl, while several cats, all completely anæstheticised, were bound down on the boards around, awaiting the healing of their wounds after the removal of tumours. At the sight of such apparent tortures the people grew wild with rage, and happening in their transports to fling down a large bottle of the anegeiric, or reviving mixture, the child and the animals all at once recovered consciousness, and began of course to writhe and scream with acute pain. I need not describe to you the scene that ensued. My laboratory was wrecked, my assistant severely injured, and I myself barely escaped with my life.

"After this contretemps I determined to be more cautious. I took the lease of a new house at Hampstead, and in the garden I determined to build myself a subterranean laboratory where I might be absolutely free from intrusion. I hired some labourers from Bath for this purpose, and I explained to them the nature of my wishes, and the absolute necessity of secrecy. A high wall surrounded the garden, and here the workmen worked securely and unseen. I concealed my design even from my dear brother—whose grandson or great-grandson I suppose you must be—and when the building was finished, I sent my men back to Bath, with strict injunctions never to mention the matter to any one. A trap-door in the cellar, artfully concealed, gave access to the passage; a large oak portal, bound with iron, shut me securely in; and my air supply was obtained by means of pipes communicating through blank spaces in the brick wall of the garden with the outer atmosphere. Every arrangement for concealment was perfect; and I resolved in future, till my results were perfectly established, that I would dispense with the aid of an assistant.

"I was in high spirits when I went to visit my Amelia that evening, and I told her confidently that before the end of the year I expected to gain the gold medal of the Royal Society. The dear girl was pleased at my glowing prospects, and gave me every assurance of the delight with which she hailed the probability of our approaching union.

"Next day I began my experiments afresh in my new quarters. I bolted myself into the laboratory, and set to work with renewed vigour. I was experimenting upon an injured dog, and I placed a large bottle of Pausodyne beside me as I administered the drug to his nostrils. The rising fumes seemed to affect my head more than usual in that confined space, and I tottered a little as I worked. My arm grew weaker, and at last fell powerless to my side. As it fell it knocked down the large bottle of Pausodyne, and I saw the liquid spreading over the floor. That was almost the last thing that I knew. I staggered toward the door, but did not reach it; and then I remember nothing more for a considerable period."

He wiped his forehead with his sleeve—he had no handkerchief—and then proceeded.

"When I woke up again the effects of the Pausodyne had worn themselves out, and I felt that I must have remained unconscious for at least a week or a fortnight. My candle had gone out, and I could not find my tinder-box. I rose up slowly and with difficulty, for the air of the room was close and filled with fumes, and made my way in the dark towards the door. To my surprise, the bolt was so stiff with rust that it would hardly move. I opened it after a struggle, and found myself in the passage. Groping my way towards the trap-door of the cellar, I felt it was obstructed by some heavy body. With an immense effort, for my strength seemed but feeble, I pushed it up, and discovered that a heap of sea-coals lay on top of it. I extricated myself into the cellar, and there a fresh surprise awaited me. A new entrance had been made into the front, so that I walked out at once upon the open road, instead of up the stairs into the kitchen. Looking up at the exterior of my house, my brain reeled with bewilderment when I saw that it had disappeared almost entirely, and that a different porch and wholly unfamiliar windows occupied its façade. I must have slept far longer than I at first imagined—perhaps a whole year or more. A vague terror prevented me from walking up the steps of my own home. Possibly my brother, thinking me dead, might have sold the lease; possibly some stranger might resent my intrusion into the house that was now his own. At any rate, I thought it safer to walk into the road. I would go towards London, to my brother's house in St. Mary le Bone. I turned into the Hampstead Road, and directed my steps thitherward.

"Again, another surprise began to affect me with a horrible and ill-defined sense of awe. Not a single object that I saw was really familiar to me. I recognized that I was in the Hampstead Road, but it was not the Hampstead Road which I used to know before my fatal experiments. The houses were far more numerous, the trees were bigger and older. A year, nay, even a few years would not have sufficed for such a change. I began to fear that I had slept away a whole decade.

"It was early morning, and few people were yet abroad. But the costume of those whom I met seemed strange and fantastic to me. Moreover, I noticed that they all turned and looked after me with evident surprise, as though my dress caused them quite as much astonishment as theirs caused me. I was quietly attired in my snuff-coloured suit of small-clothes, with silk stockings and simple buckle shoes, and I had of course no hat; but I gathered that my appearance caused universal amazement and concern, far more than could be justified by the mere accidental absence of head-gear. A dread began to oppress me that I might actually have slept out my whole age and generation. Was my Amelia alive? and if so, would she be still the same Amelia I had known a week or two before? Should I find her an aged woman, still cherishing a reminiscence of her former love; or might she herself perhaps be dead and forgotten, while I remained, alone and solitary, in a world which knew me not?

"I walked along unmolested, but with reeling brain, through streets more and more unfamiliar, till I came near the St. Mary le Bone Road. There, as I hesitated a little and staggered at the crossing, a man in a curious suit of dark blue clothes, with a grotesque felt helmet on his head, whom I afterwards found to be a constable, came up and touched me on the shoulder.

"'Look here,' he said to me in a rough voice, 'what are you a-doin' in this 'ere fancy-dress at this hour in the mornin'? You've lost your way home, I take it.'

"'I was going,' I answered, 'to the St. Mary le Bone Road.'

"'Why, you image,' says he rudely, 'if you mean Marribon, why don't you say Marribon? What house are you a-lookin' for, eh?'

"'My brother lives,' I replied, 'at the Lamb, near St. Mary's Church, and I was going to his residence.'

"'The Lamb!' says he, with a rude laugh; 'there ain't no public of that name in the road. It's my belief,' he goes on after a moment, 'that you're drunk, or mad, or else you've stole them clothes. Any way, you've got to go along with me to the station, so walk it, will you?'

"'Pardon me,' I said, 'I suppose you are an officer of the law, and I would not attempt to resist your authority'—'You'd better not,' says he, half to himself—'but I should like to go to my brother's house, where I could show you that I am a respectable person.'

"'Well,' says my fellow insolently, 'I'll go along of you if you like, and if it's all right, I suppose you won't mind standing a bob?'

"'A what?' said I.

"'A bob,' says he, laughing; 'a shillin', you know.'

"To get rid of his insolence for a while, I pulled out my purse and handed him a shilling. It was a George II with milled edges, not like the things I see you use now. He held it up and looked at it, and then he said again, 'Look here, you know, this isn't good. You'd better come along with me straight to the station, and not make a fuss about it. There's three charges against you, that's all. One is, that you're drunk. The second is, that you're mad. And the third is, that you've been trying to utter false coin. Any one of 'em's quite enough to justify me in takin' you into custody.'

"I saw it was no use to resist, and I went along with him.

"I won't trouble you with the whole of the details, but the upshot of it all was, they took me before a magistrate. By this time I had begun to realize the full terror of the situation, and I saw clearly that the real danger lay in the inevitable suspicion of madness under which I must labour. When I got into the court I told the magistrate my story very shortly and simply, as I have told it to you now. He listened to me without a word, and at the end he turned round to his clerk and said, 'This is clearly a case for Dr. Fitz-Jenkins, I think.'

"'Sir,' I said, 'before you send me to a madhouse, which I suppose is what you mean by these words, I trust you will at least examine the evidences of my story. Look at my clothing, look at these coins, look at everything about me.' And I handed him my purse to see for himself.

"He looked at it for a minute, and then he turned towards me very sternly. 'Mr. Spottiswood,' he said, 'or whatever else your real name may be, if this is a joke, it is a very foolish and unbecoming one. Your dress is no doubt very well designed; your small collection of coins is interesting and well-selected; and you have got up your character remarkably well. If you are really sane, which I suspect to be the case, then your studied attempt to waste the time of this court and to make a laughing-stock of its magistrate will meet with the punishment it deserves. I shall remit your case for consideration to our medical officer. If you consent to give him your real name and address, you will be liberated after his examination. Otherwise, it will be necessary to satisfy ourselves as to your identity. Not a word more, sir,' he continued, as I tried to speak on behalf of my story. 'Inspector, remove the prisoner.'

"They took me away, and the surgeon examined me. To cut things short, I was pronounced mad, and three days later the commissioners passed me for a pauper asylum. When I came to be examined, they said I showed no recollection of most subjects of ordinary education.

"'I am a chemist,' said I; 'try me with some chemical questions. You will see that I can answer sanely enough.'

"'How do you mix a grey powder?' said the commissioner.

"'Excuse me,' I said, 'I mean a chemical philosopher, not an apothecary.'

"'Oh, very well, then; what is carbonic acid?'

"'I never heard of it,' I answered in despair. 'It must be something which has come into use since—since I left off learning chemistry.' For I had discovered that my only chance now was to avoid all reference to my past life and the extraordinary calamity which had thus unexpectedly overtaken me. 'Please try me with something else.'

"'Oh, certainly. What is the atomic weight of chlorine?'

"I could only answer that I did not know.

"'This is a very clear case,' said the commissioner. 'Evidently he is a gentleman by birth and education, but he can give no very satisfactory account of his friends, and till they come forward to claim him we can only send him for a time to North Street.'

"'For Heaven's sake, gentlemen,' I cried, 'before you consign me to an asylum, give me one more chance. I am perfectly sane; I remember all I ever knew; but you are asking me questions about subjects on which I never had any information. Ask me anything historical, and see whether I have forgotten or confused any of my facts.'

"I will do the commissioner the justice to say that he seemed anxious not to decide upon the case without full consideration. 'Tell me what you can recollect,' he said, 'as to the reign of George IV.'

"'I know nothing at all about it,' I answered, terror-stricken, 'but oh, do pray ask me anything up to the time of George III.'

"'Then please say what you think of the French Revolution.'

"I was thunderstruck. I could make no reply, and the commissioners shortly signed the papers to send me to North Street pauper asylum. They hurried me into the street, and I walked beside my captors towards the prison to which they had consigned me. Yet I did not give up all hope even so of ultimately regaining my freedom. I thought the rationality of my demeanour and the obvious soundness of all my reasoning powers would suffice in time to satisfy the medical attendant as to my perfect sanity. I felt sure that people could never long mistake a man so clear-headed and collected as myself for a madman.

"On our way, however, we happened to pass a churchyard where some workmen were engaged in removing a number of old tombstones from the crowded area. Even in my existing agitated condition, I could not help catching the name and date on one mouldering slab which a labourer had just placed upon the edge of the pavement. It ran something like this: 'Sacred to the memory of Amelia, second daughter of the late Sir Thomas Spragg, knight, and beloved wife of Henry McAlister, Esq., by whom this stone is erected. Died May 20, 1799, aged 44 years.' Though I had gathered already that my dear girl must probably have long been dead, yet the reality of the fact had not yet had time to fix itself upon my mind. You must remember, my dear sir, that I had but awaked a few days earlier from my long slumber, and that during those days I had been harassed and agitated by such a flood of incomprehensible complications, that I could not really grasp in all its fulness the complete isolation of my present position. When I saw the tombstone of one whom, as it seemed to me, I had loved passionately but a week or two before, I could not refrain from rushing to embrace it, and covering the insensible stone with my boiling tears. 'Oh, my Amelia, my Amelia,' I cried, 'I shall never again behold thee, then! I shall never again press thee to my heart, or hear thy dear lips pronounce my name!'

"But the unfeeling wretches who had charge of me were far from being moved to sympathy by my bitter grief. 'Died in 1799,' said one of them with a sneer. 'Why, this madman's blubbering over the grave of an old lady who has been buried for about a hundred years!' And the workmen joined in their laughter as my gaolers tore me away to the prison where I was to spend the remainder of my days.

"When we arrived at the asylum, the surgeon in attendance was informed of this circumstance, and the opinion that I was hopelessly mad thus became ingrained in his whole conceptions of my case. I remained five months or more in the asylum, but I never saw any chance of creating a more favourable impression on the minds of the authorities. Mixing as I did only with other patients, I could gain no clear ideas of what had happened since I had taken my fatal sleep; and whenever I endeavoured to question the keepers, they amused themselves by giving me evidently false and inconsistent answers, in order to enjoy my chagrin and confusion. I could not even learn the actual date of the present year, for one keeper would laugh and say it was 2001, while another would confidentially advise me to date my petition to the Commissioners, 'Jan. 1, A.D. one million.' The surgeon, who never played me any such pranks, yet refused to aid me in any way, lest, as he said, he should strengthen me in my sad delusion. He was convinced that I must be an historical student, whose reason had broken down through too close study of the eighteenth century; and he felt certain that sooner or later my friends would come to claim me. He is a gentle and humane man, against whom I have no personal complaint to make; but his initial misconception prevented him and everybody else from ever paying the least attention to my story. I could not even induce them to make inquiries at my house at Hampstead, where the discovery of the subterranean laboratory would have partially proved the truth of my account.

"Many visitors came to the asylum from time to time, and they were always told that I possessed a minute and remarkable acquaintance with the history of the eighteenth century. They questioned me about facts which are as vivid in my memory as those of the present month, and were much surprised at the accuracy of my replies. But they only thought it strange that so clever a man should be so very mad, and that my information should be so full as to past events, while my notions about the modern world were so utterly chaotic. The surgeon, however, always believed that my reticence about all events posterior to 1781 was a part of my insanity. I had studied the early part of the eighteenth century so fully, he said, that I fancied I had lived in it; and I had persuaded myself that I knew nothing at all about the subsequent state of the world."

The poor fellow stopped a while, and again drew his sleeve across his forehead. It was impossible to look at him and believe for a moment that he was a madman.

"And how did you make your escape from the asylum?" I asked.

"Now, this very evening," he answered; "I simply broke away from the door and ran down toward the Strand, till I came to a place that looked a little like St. Martin's Fields, with a great column and some fountains, and near there I met you. It seemed to me that the best thing to do was to catch the York coach and get away from the town as soon as possible. You met me, and your look and name inspired me with confidence. I believe you must be a descendant of my dear brother."

"I have not the slightest doubt," I answered solemnly, "that every word of your story is true, and that you are really my great-great-uncle. My own knowledge of our family history exactly tallies with what you tell me. I shall spare no endeavour to clear up this extraordinary matter, and to put you once more in your true position."

"And you will protect me?" he cried fervently, clasping my hand in both his own with intense eagerness. "You will not give me up once more to the asylum people?"

"I will do everything on earth that is possible for you," I replied.

He lifted my hand to his lips and kissed it several times, while I felt hot tears falling upon it as he bent over me. It was a strange position, look at it how you will. Grant that I was but the dupe of a madman, yet even to believe for a moment that I, a man of well-nigh fifty, stood there in face of my own great-grandfather's brother, to all appearance some twenty years my junior, was in itself an extraordinary and marvellous thing. Both of us were too overcome to speak. It was a few minutes before we said anything, and then a loud knock at the door made my hunted stranger rise up hastily in terror from his chair.

"Gracious Heavens!" he cried, "they have tracked me hither. They are coming to fetch me. Oh, hide me, hide me, anywhere from these wretches!"

As he spoke, the door opened, and two keepers with a policeman entered my room.

"Ah, here he is!" said one of them, advancing towards the fugitive, who shrank away towards the window as he approached.

"Do not touch him," I exclaimed, throwing myself in the way. "Every word of what he says is true, and he is no more insane than I am."

The keeper laughed a low laugh of vulgar incredulity. "Why, there's a pair of you, I do believe," he said. "You're just as mad yourself as t'other one." And he pushed me aside roughly to get at his charge.

But the poor fellow, seeing him come towards him, seemed suddenly to grow instinct with a terrible vigour, and hurled off the keeper with one hand, as a strong man might do with a little terrier. Then, before we could see what he was meditating, he jumped upon the ledge of the open window, shouted out loudly, "Farewell, farewell!" and leapt with a spring on to the embankment beneath.

All four of us rushed hastily down the three flights of steps to the bottom, and came below upon a crushed and mangled mass on the spattered pavement. He was quite dead. Even the policeman was shocked and horrified at the dreadful way in which the body had been crushed and mutilated in its fall, and at the suddenness and unexpectedness of the tragedy. We took him up and laid him out in my room; and from that room he was interred after the inquest, with all the respect which I should have paid to an undoubted relative. On his grave in Kensal Green Cemetery I have placed a stone bearing the simple inscription, "Jonathan Spottiswood. Died 1881." The hint I had received from the keeper prevented me from saying anything as to my belief in his story, but I asked for leave to undertake the duty of his interment on the ground that he bore my own surname, and that no other person was forthcoming to assume the task. The parochial authorities were glad enough to rid the ratepayers of the expense.

At the inquest I gave my evidence simply and briefly, dwelling mainly upon the accidental nature of our meeting, and the facts as to his fatal leap. I said nothing about the known disappearance of Jonathan Spottiswood in 1781, nor the other points which gave credibility to his strange tale. But from this day forward I give myself up to proving the truth of his story, and realizing the splendid chemical discovery which promises so much benefit to mankind. For the first purpose, I have offered a large reward for the discovery of a trap-door in a coal-cellar at Hampstead, leading into a subterranean passage and laboratory; since, unfortunately, my unhappy visitor did not happen to mention the position of his house. For the second purpose, I have begun a series of experiments upon the properties of the essential oil of alchemilla, and the possibility of successfully treating it with carbonic anhydride; since, unfortunately, he was equally vague as to the nature of his process and the proportions of either constituent. Many people will conclude at once, no doubt, that I myself have become infected with the monomania of my miserable namesake, but I am determined at any rate not to allow so extraordinary an anæsthetic to go unacknowledged, if there be even a remote chance of actually proving its useful nature. Meanwhile, I say nothing even to my dearest friends with regard to the researches upon which I am engaged.

> **Charles Grant Blairfindie Allen** (1848–1899) was a Canadian-born British author who wrote under the name Grant Allen and also, as here, J. Arbuthnot Wilson. He wrote nonfiction books about evolution and Darwinism, and a number of science fiction short stories. He was a friend of H. G. Wells, and the influence of the great writer can be spotted in several of Allen's stories.

Commentary

The notion of suspended animation has a long history in science fiction, and for good reason. This literary device gives writers a way to move their characters into the future. Most stories, though, pay scant attention to the science of putting humans into a hibernation state. At least Allen, writing as Wilson in this 1881 story, attempts a scientific rationale. Allen has his protagonist make a study of "fixed air". The protagonist's interlocutor points out that fixed air is "the old-fashioned name for carbonic acid gas". In turn I should point out that carbonic acid gas is an old-fashioned name for carbon dioxide, CO_2. In other words, Allen's protagonist discovers that Pausodyne, a CO_2-containing compound, induces a state of suspended animation.

Allen's suggestion, although not particularly plausible, at least gives a nod to the known science of the time. People knew they became drowsy when they breathed air containing too much CO_2. And in old coal mines, blackdamp—a mix of carbon dioxide and nitrogen—could render miners unconscious. For the Victorian SF writer wanting to rationalise the properties of a substance such as Pausodyne, it made sense to mention CO_2.

The modern SF writer wanting to rationalise the concept of suspended animation often settles upon cryonics. Cryonics relies on freezing and storing a human body, with the hope of future resurrection. And people have already attempted to apply the technique. (Well, people have applied the freezing and storage part. The resurrection part remains a distant ambition.) The first patient/customer for the technique? James Bedford, who died on 12 January 1967. Within two hours of his death chemists had cryopreserved his corpse in a vat of liquid nitrogen. Since then, hundreds of other people have had their dead bodies frozen and stored at a temperature of 77.1 K, or $-196\ °C$ (the boiling point of liquid nitrogen).

The idea of bringing the dead back to life seems crazy. But is it?

Biologists have long known they can reanimate primitive organisms that have survived for decades in a dormant state. But the tenacity with which life clings on astonishes even biologists.

Consider life in the deep subseafloor, a place where microbes get nutrients from the organic detritus falling from the upper reaches of the ocean. The South Pacific Gyre, far from land, remote from productive volumes of ocean, and isolated by continuous circular currents and winds, forms a 'desert' in the ocean, a spot with low nutrient levels on the seafloor. In 2020, scientists drilled deep into the seabed beneath the Gyre and brought sediment cores to the surface. The biologists did not expect to find microbes in samples of sediment laid down over 100 million years ago. But find microbes they did. And when revived, the microbes multiplied in the laboratory.

Or consider that wonderful creature the tardigrade, also called the water bear (see Fig. 10.1). A tardigrade is nigh on indestructible. When it encounters a harsh environment (heat, pressure, radiation, drought) it enters a dormant state. Its metabolism drops to less than 0.01% of normal and its water

Fig. 10.1 An image of a tardigrade (also known as a water bear) taken by a scanning electron microscope. The image shows the tardigrade's barrel-shaped body and its four pairs of short legs. A typical tardigrade grows to 0.3–0.5 mm so, in the right light, you can just about see them with the unaided eye. When frozen, tardigrades enter a state resembling a deep sleep; upon thawing, they carry on as normal. (They can also survive extremes of heat, vacuum, and radiation.) (Credit: Schokraie E., Warnken U., Hotz-Wagenblatt A., Grohme M.A., Hengherr S., et al., CC BY 2.5 DEED)

content to 1% of normal. When its environment improves, the tardigrade wakes up.

Microbes and tardigrades exemplify simple forms of life, but even sophisticated creatures can hibernate. Organisms from butterflies to bats turn down their metabolism during cold seasons. Hibernation allows them to save energy. In hot climates, some creatures—Nile crocodiles, for example—escape heat or drought by using a similar trick. In this case scientists call the process aestivation rather than hibernation, but the effect remains the same.

Humans, of course, are more complicated creatures than microbes or tardigrades. We cannot hibernate like a hedgehog. We cannot aestivate like an earthworm. But maybe we can have our animation suspended? In 1999, newspapers reported the case of Anna Bågenholm, a Swedish medic. She found herself trapped under the ice on a frozen lake for 80 minutes. Her heart stopped, but she survived without brain damage. In 2015, something similar happened to a 14-year-old American boy called John Smith. He survived after spending 15 minutes under a frozen lake. And in 2001, Canadian medics treated the 13-month-old Erika Nordby. The toddler had stepped out of her house alone, almost naked, on a bitterly cold night. For two hours, she had no detectable heartbeat and a body temperature of just 16 °C. Medics revived her.

In principle, then, cryonics-based suspended animation seems possible.

That tantalising possibility means some people will pay to have their bodies cryogenically preserved after death. Others will take their money and provide the service. Obstacles, of course, exist.

One obvious obstacle: funding. You have to pay for cryopreservation. Cryonics providers charge to store your body or head in the dewar of liquid nitrogen (see Fig. 10.2) that keeps your tissue frozen at −196 °C. Liquid nitrogen tends to boil away, though, so workers must ensure the dewar remains topped up. That costs. Indeed, some pioneering cryonics providers have gone out of business. In those cases the authorities thawed the customers and disposed of the corpses in a conventional manner. Today, cryonics providers typically charge an annual membership fee. Upon the death of the member they ask for a large lump sum on top, a fee that typically comes from life insurance, and place the money in a trust fund to pay for ongoing storage costs.

Jurisprudence can create another obstacle. Some countries permit just three legal ways to dispose of a body: burial, cremation, and donation to science. If you pay for cryonics, make sure you die in the right country.

Yet another obstacle, to my mind, involves the hubris attached to the idea of reanimation. Why would people in the distant future feel any obligation to bring a corpse back to life? If future civilisations want to bring more life into

Fig. 10.2 A feeding tube fills a dewar with liquid nitrogen, at the Warm Springs Fish Technology Center in Georgia. The Center uses cryopreservation techniques to freeze and store living cells. If those cells remain viable for future growth upon thawing, these techniques might help secure the genetic diversity of endangered species. When in storage the samples must remain frozen, which means technicians must top up the dewars once a week. In commercial cryonics organisations, similar sized dewars store the brains of deceased humans who hope for reanimation. Full-body storage of course requires larger dewars. (Credit: Cody Meshes, USFWS, CC BY 2.0 DEED)

their world then they can encourage parenthood. Babies bring with them new possibilities. Why reanimate someone who has already lived a life? (Using that person as an object of study, though not a selling point for the cryonics industry, would I suppose constitute one reason for reanimation.)

And technical obstacles abound. At present, medical science cannot recover large animals after freezing. The major organs tend to fracture during cooling. As for a complex organ such as the human brain—well, who knows what damage it sustains during cooling? (The brains currently under cryopreservation dwell inside locked canisters of liquid nitrogen. No one knows what goes on with them.) In fairness, proponents of cryonics understand the required technology does not exist. They hope the technology will exist at some future date.

A final obstacle. Let's assume the reanimation of the frozen brain becomes possible one day. When doctors reanimate a brain, which individual will they return to life? I know of an elderly person, a long-time member of a cryonics organisation, who, 18 months ago as I write, suffered a severe, debilitating stroke. According to close relatives, the stroke took away something vital. One day that person will die and, I presume, go into cryopreservation. Upon reanimation, would that person continue to suffer the effects of stroke? One hopes not. I presume the patient wants future doctors to somehow excise the effects of those 18 months. The patient wishes for a reanimated brain that functions as it did before the stroke. But why stop there? If future doctors can repair the damaged brain of an 80-year-old and make it that of a healthy 78-year-old, why not make it that of a 50-year-old? Or a 20-year-old? In that case, though, in what sense is the reanimated person the same as the individual who died?

Several organisations offer a cryonics service. The Alcor Life Extension Foundation, one of the best known companies in the field, has over 1300 members. Alcor allows international membership, but its services remain limited outside its base in America. The Cryonics Institute, also based in America, has more than 1100 members across the globe. Tomorrow Biostasis, based in Germany, has more than 250 members. Other organisations operate in Russia and China. Given the obstacles outlined above, though, why do so many people willingly hand over wads of money to these organisations?

Enthusiasts perhaps look on cryonics as a form of Pascal's wager. Blaise Pascal argued that you may as well believe in God and live that way. If God does not exist, then you have lost nothing. (Except the chance for some fun, I suppose, depending on how you think God wants you to behave.) And if God does exist, then you gain everything. Personally, I find Simon Dein's take on cryonics telling. Dein, a consultant psychiatrist, argues that the field of cryonics rests on assertions that science has not tested or established, and potentially those assertions will always remain out of reach of empirical demonstration. Cryonics places naive faith in non-existent technology and promises to overcome death: one sees more religion here than science. In Allen's tale, the unfortunate Jonathan Spottiswood could not cheat death. He did, though, have his animation suspended. I doubt the modern-day Spottiswoods will achieve even that level of success.

Notes and Further Reading

first patient/customer for the technique—For the story of Bedford's cryonic suspension, see Perry (1991).

scientists drilled deep into the seabed—A non-technical account of the discovery of living bacteria in 100-million-year-old sediment can be found in Morono (2021). The technical paper of the same discovery is Morono et al. (2020).

that wonderful creature the tardigrade—Ingemar Jönsson et al. (2008) describe how some tardigrades have survived combined exposure to the vacuum of space and solar radiation.

the case of Anna Bågenholm—The treatment that led to Anna Bågenholm's recovery is reported by Gilbert et al. (2000). In 2019, John Smith's story was made into the feature film *Breakthrough*, distributed by twentieth century Fox. In 2002, Erika Nordby's experience was the focus of a television show on Discovery Health Channel.

organisations offer a cryonics service—For a review of cryonics companies active as of late 2022, see GrowingLife (2022).

Dein's take on cryonics—Professor Simon Dean writes widely on religion and health. For his argument regarding cryonics, see Dean (2022).

References

Dean, S.: Cryonics: science or religion? J. Relig. Health. **61**(4), 3164–3176 (2022). https://doi.org/10.1007/s10943-020-01166-6

Gilbert, M., Busund, R., Skagseth, A., Nilsen, P.A., Solbø, J.P.: Resuscitation from accidental hypothermia of 13.7 °C with circulatory arrest. Lancet. **355**(9201), P375–6. 164–76 (2000). https://doi.org/10.1016/S0140-6736(00)01021-7

GrowingLife: A tour of the companies seeking to revive the dead? www.gowinglife.com/a-tour-of-the-companies-seeking-to-revive-the-dead/ (2022)

Ingemar Jönsson, K., et al.: Tardigrades survive exposure to space in low Earth orbit. Curr. Biol. **18**(17), R729–R731 (2008). https://doi.org/10.1016/j.cub.2008.06.048

Morono, Y.: Buried alive. The Biologist. www.rsb.org.uk/biologist-features/buried-alive (2021)

Morono, Y., et al.: Aerobic microbial life persists in oxic marine sediment as old as 101.5 million years. Nat. Commun. **11**, 3626 (2020)

Perry, M.: The first suspension. Cryonics. www.cryonicsarchive.org/library/bedford-suspension/ (1991)

11

Advances in Medical Technology

The Facts in the Case of M. Valdemar (Edgar Allan Poe)

Of course I shall not pretend to consider it any matter for wonder, that the extraordinary case of M. Valdemar has excited discussion. It would have been a miracle had it not—especially under the circumstances. Through the desire of all parties concerned, to keep the affair from the public, at least for the present, or until we had farther opportunities for investigation—through our endeavours to effect this—a garbled or exaggerated account made its way into society, and became the source of many unpleasant misrepresentations; and, very naturally, of a great deal of disbelief.

It is now rendered necessary that I give the facts—as far as I comprehend them myself. They are, succinctly, these:

My attention, for the last three years, had been repeatedly drawn to the subject of Mesmerism; and, about nine months ago, it occurred to me, quite suddenly, that in the series of experiments made hitherto, there had been a very remarkable and most unaccountable omission:—no person had as yet been mesmerized in articulo mortis. It remained to be seen, first, whether, in such condition, there existed in the patient any susceptibility to the magnetic influence; secondly, whether, if any existed, it was impaired or increased by the condition; thirdly, to what extent, or for how long a period, the encroachments of Death might be arrested by the process. There were other points to be ascertained, but these most excited my curiosity—the last in especial, from the immensely important character of its consequences.

In looking around me for some subject by whose means I might test these particulars, I was brought to think of my friend, M. Ernest Valdemar, the well-known compiler of the "Bibliotheca Forensica", and author (under the nom de plume of Issachar Marx) of the Polish versions of "Wallenstein" and "Gargantua". M. Valdemar, who has resided principally at Harlem, N.Y., since the year 1839, is (or was) particularly noticeable for the extreme spareness of his person—his lower limbs much resembling those of John Randolph; and, also, for the whiteness of his whiskers, in violent contrast to the blackness of his hair—the latter, in consequence, being very generally mistaken for a wig. His temperament was markedly nervous, and rendered him a good subject for mesmeric experiment. On two or three occasions I had put him to sleep with little difficulty, but was disappointed in other results which his peculiar constitution had naturally led me to anticipate. His will was at no period positively, or thoroughly, under my control, and in regard to clairvoyance, I could accomplish with him nothing to be relied upon. I always attributed my failure at these points to the disordered state of his health. For some months previous to my becoming acquainted with him, his physicians had declared him in a confirmed phthisis. It was his custom, indeed, to speak calmly of his approaching dissolution, as of a matter neither to be avoided nor regretted.

When the ideas to which I have alluded first occurred to me, it was of course very natural that I should think of M. Valdemar. I knew the steady philosophy of the man too well to apprehend any scruples from him; and he had no relatives in America who would be likely to interfere. I spoke to him frankly upon the subject; and, to my surprise, his interest seemed vividly excited. I say to my surprise; for, although he had always yielded his person freely to my experiments, he had never before given me any tokens of sympathy with what I did. His disease was of that character which would admit of exact calculation in respect to the epoch of its termination in death; and it was finally arranged between us that he would send for me about twenty-four hours before the period announced by his physicians as that of his decease.

It is now rather more than 7 months since I received, from M. Valdemar himself, the subjoined note:

MY DEAR P——,

You may as well come now. D—— and F—— are agreed that I cannot hold out beyond tomorrow midnight; and I think they have hit the time very nearly.

VALDEMAR.

I received this note within half an hour after it was written, and in fifteen minutes more I was in the dying man's chamber. I had not seen him for ten days, and was appalled by the fearful alteration which the brief interval had wrought in him. His face wore a leaden hue; the eyes were utterly lustreless; and the emaciation was so extreme, that the skin had been broken through by the cheek-bones. His expectoration was excessive. The pulse was barely perceptible. He retained, nevertheless, in a very remarkable manner, both his mental power and a certain degree of physical strength. He spoke with distinctness—took some palliative medicines without aid—and, when I entered the room, was occupied in penciling memoranda in a pocket-book. He was propped up in the bed by pillows. Doctors D—— and F—— were in attendance.

After pressing Valdemar's hand, I took these gentlemen aside, and obtained from them a minute account of the patient's condition. The left lung had been for 18 months in a semi-osseous or cartilaginous state, and was, of course, entirely useless for all purposes of vitality. The right, in its upper portion, was also partially, if not thoroughly, ossified, while the lower region was merely a mass of purulent tubercles, running one into another. Several extensive perforations existed; and, at one point, permanent adhesion to the ribs had taken place. These appearances in the right lobe were of comparatively recent date. The ossification had proceeded with very unusual rapidity; no sign of it had been discovered a month before, and the adhesion had only been observed during the three previous days. Independently of the phthisis, the patient was suspected of aneurism of the aorta; but on this point the osseous symptoms rendered an exact diagnosis impossible. It was the opinion of both physicians that M. Valdemar would die about midnight on the morrow (Sunday). It was then seven o'clock on Saturday evening.

On quitting the invalid's bed-side to hold conversation with myself, Doctors D—— and F—— had bidden him a final farewell. It had not been their intention to return; but, at my request, they agreed to look in upon the patient about ten the next night.

When they had gone, I spoke freely with M. Valdemar on the subject of his approaching dissolution, as well as, more particularly, of the experiment proposed. He still professed himself quite willing and even anxious to have it made, and urged me to commence it at once. A male and a female nurse were in attendance; but I did not feel myself altogether at liberty to engage in a task of this character with no more reliable witnesses than these people, in case of sudden accident, might prove. I therefore postponed operations until about eight the next night, when the arrival of a medical student, with whom I had some acquaintance, (Mr Theodore L——l,) relieved me from farther embarrassment. It had been my design, originally, to wait for the physicians; but I was induced to proceed, first, by the urgent entreaties of M. Valdemar, and secondly, by my conviction that I had not a moment to lose, as he was evidently sinking fast.

Mr. L——l was so kind as to accede to my desire that he would take notes of all that occurred; and it is from his memoranda that what I now have to relate is, for the most part, either condensed or copied verbatim.

It wanted about five minutes of eight when, taking the patient's hand, I begged him to state, as distinctly as he could, to Mr. L——l, whether he (M. Valdemar,) was entirely willing that I should make the experiment of mesmerizing him in his then condition.

He replied feebly, yet quite audibly, "Yes, I wish to be mesmerized"—adding immediately afterwards, "I fear you have deferred it too long."

While he spoke thus, I commenced the passes which I had already found most effectual in subduing him. He was evidently influenced with the first lateral stroke of my hand across his forehead; but although I exerted all my powers, no farther perceptible effect was induced until some minutes after ten o'clock, when Doctors D—— and F—— called, according to appointment. I explained to them, in a few words, what I designed, and as they opposed no objection, saying that the patient was already in the death agony, I proceeded without hesitation—exchanging, however, the lateral passes for downward ones, and directing my gaze entirely into the right eye of the sufferer.

By this time his pulse was imperceptible and his breathing was stertorous, and at intervals of half a minute.

This condition was nearly unaltered for a quarter of an hour. At the expiration of this period, however, a natural although a very deep sigh escaped the bosom of the dying man, and the stertorous breathing ceased—that is to say, its stertorousness was no longer apparent; the intervals were undiminished. The patient's extremities were of an icy coldness.

At five minutes before eleven, I perceived unequivocal signs of the mesmeric influence. The glassy roll of the eye was changed for that expression of uneasy inward examination which is never seen except in cases of sleep-waking, and which it is quite impossible to mistake. With a few rapid lateral passes I made the lids quiver, as in incipient sleep, and with a few more I closed them altogether. I was not satisfied, however, with this, but continued the manipulations vigorously, and with the fullest exertion of the will, until I had completely stiffened the limbs of the slumberer, after placing them in a seemingly easy position. The legs were at full length; the arms were nearly so, and reposed on the bed at a moderate distance from the loins. The head was very slightly elevated.

When I had accomplished this, it was fully midnight, and I requested the gentlemen present to examine M. Valdemar's condition. After a few experiments, they admitted him to be in an unusually perfect state of mesmeric trance. The curiosity of both the physicians was greatly excited. Dr. D—— resolved at once to remain with the patient all night, while Dr. F—— took leave with a promise to return at daybreak. Mr. L——l and the nurses remained.

We left M. Valdemar entirely undisturbed until about three o'clock in the morning, when I approached him and found him in precisely the same condition as when Dr. F—— went away—that is to say, he lay in the same position; the pulse was imperceptible; the breathing was gentle (scarcely noticeable, unless through the application of a mirror to the lips;) the eyes were closed naturally; and the limbs were as rigid and as cold as marble. Still, the general appearance was certainly not that of death.

As I approached M. Valdemar I made a kind of half effort to influence his right arm into pursuit of my own, as I passed the latter gently to and fro above his person. In such experiments with this patient, I had never perfectly succeeded before, and assuredly I had little thought of succeeding now; but to my astonishment, his arm very readily, although feebly, followed every direction I assigned it with mine. I determined to hazard a few words of conversation.

"M. Valdemar," I said, "are you asleep?" He made no answer, but I perceived a tremor about the lips, and was thus induced to repeat the question, again and again. At its third repetition, his whole frame was agitated by a very slight shivering; the eye-lids unclosed themselves so far as to display a white line of a ball; the lips moved sluggishly, and from between them, in a barely audible whisper, issued the words:

"Yes;—asleep now. Do not wake me!—let me die so!"

I here felt the limbs and found them as rigid as ever. The right arm, as before, obeyed the direction of my hand. I questioned the sleep-waker again:

"Do you still feel pain in the breast, M. Valdemar?"

The answer now was immediate, but even less audible than before:

"No pain—I am dying."

I did not think it advisable to disturb him farther just then, and nothing more was said or done until the arrival of Dr. F——, who came a little before sunrise, and expressed unbounded astonishment at finding the patient still alive. After feeling the pulse and applying a mirror to the lips, he requested me to speak to the sleep-waker again. I did so, saying:

"M. Valdemar, do you still sleep?"

As before, some minutes elapsed ere a reply was made; and during the interval the dying man seemed to be collecting his energies to speak. At my fourth repetition of the question, he said very faintly, almost inaudibly:

"Yes; still asleep—dying."

It was now the opinion, or rather the wish, of the physicians, that M. Valdemar should be suffered to remain undisturbed in his present apparently tranquil condition, until death should supervene—and this, it was generally agreed, must now take place within a few minutes. I concluded, however, to speak to him once more, and merely repeated my previous question.

While I spoke, there came a marked change over the countenance of the sleep-waker. The eyes rolled themselves slowly open, the pupils disappearing upwardly; the skin generally assumed a cadaverous hue, resembling not so much parchment as white paper; and the circular hectic spots which, hitherto, had been strongly defined in the centre of each cheek, went out at once. I use this expression, because the suddenness of their departure put me in mind of nothing so much as the extinguishment of a candle by a puff of the breath. The upper lip, at the same time, writhed itself away from the teeth, which it had previously covered completely; while the lower jaw fell with an audible jerk, leaving the mouth widely extended, and disclosing in full view the swollen and blackened tongue. I presume that no member of the party then present had been unaccustomed to death-bed horrors; but so hideous beyond conception was the appearance of M. Valdemar at this moment, that there was a general shrinking back from the region of the bed.

I now feel that I have reached a point of this narrative at which every reader will be startled into positive disbelief. It is my business, however, simply to proceed.

There was no longer the faintest sign of vitality in M. Valdemar; and concluding him to be dead, we were consigning him to the charge of the nurses, when a strong vibratory motion was observable in the tongue. This continued for perhaps a minute. At the expiration of this period, there issued from the distended and motionless jaws a voice—such as it would be madness in me to attempt describing. There are, indeed, two or three epithets which might be considered as applicable to it in part; I might say, for example, that the sound was harsh, and broken and hollow; but the hideous whole is indescribable, for the simple reason that no similar sounds have ever jarred upon the ear of humanity. There were two particulars, nevertheless, which I thought then, and still think, might fairly be stated as characteristic of the intonation—as well adapted to convey some idea of its unearthly peculiarity. In the first place, the voice seemed to reach our ears—at least mine—from a vast distance, or from some deep cavern within the earth. In the second place, it impressed me (I fear, indeed, that it will be impossible to make myself comprehended) as gelatinous or glutinous matters impress the sense of touch.

I have spoken both of "sound" and of "voice". I mean to say that the sound was one of distinct—of even wonderfully, thrillingly distinct—syllabification.

M. Valdemar spoke—obviously in reply to the question I had propounded to him a few minutes before. I had asked him, it will be remembered, if he still slept. He now said:

"Yes;—no;—I have been sleeping—and now—now—I am dead."

No person present even affected to deny, or attempted to repress, the unutterable, shuddering horror which these few words, thus uttered, were so well calculated to convey. Mr. L——l (the student) swooned. The nurses immediately left the chamber, and could not be induced to return. My own impressions I would not pretend to render intelligible to the reader. For nearly an hour, we busied ourselves, silently—without the utterance of a word—in endeavours to revive Mr. L——l. When he came to himself, we addressed ourselves again to an investigation of M. Valdemar's condition.

It remained in all respects as I have last described it, with the exception that the mirror no longer afforded evidence of respiration. An attempt to draw blood from the arm failed. I should mention, too, that this limb was no farther subject to my will. I endeavoured in vain to make it follow the direction of my hand. The only real indication, indeed, of the mesmeric influence, was now found in the vibratory movement of the tongue, whenever I addressed M. Valdemar a question. He seemed to be making an effort to reply, but had no longer sufficient volition. To queries put to him by any other person than myself he seemed utterly insensible—although I endeavoured to place each member of the company in mesmeric rapport with him. I believe that I have now related all that is necessary to an understanding of the sleep-waker's state at this epoch. Other nurses were procured; and at ten o'clock I left the house in company with the two physicians and Mr. L——l.

In the afternoon we all called again to see the patient. His condition remained precisely the same. We had now some discussion as to the propriety and feasibility of awakening him; but we had little difficulty in agreeing that no good purpose would be served by so doing. It was evident that, so far, death (or what is usually termed death) had been arrested by the mesmeric process. It seemed clear to us all that to awaken M. Valdemar would be merely to insure his instant, or at least his speedy dissolution.

From this period until the close of last week—an interval of nearly 7 months—we continued to make daily calls at M. Valdemar's house, accompanied, now and then, by medical and other friends. All this time the sleeper-waker remained exactly as I have last described him. The nurses' attentions were continual.

It was on Friday last that we finally resolved to make the experiment of awakening, or attempting to awaken him; and it is the (perhaps) unfortunate result of this latter experiment which has given rise to so much discussion in private circles—to so much of what I cannot help thinking unwarranted popular feeling.

For the purpose of relieving M. Valdemar from the mesmeric trance, I made use of the customary passes. These, for a time, were unsuccessful. The first indication of revival was afforded by a partial descent of the iris. It was observed, as especially remarkable, that this lowering of the pupil was accompanied by the profuse out-flowing of a yellowish ichor (from beneath the lids) of a pungent and highly offensive odor.

It was now suggested that I should attempt to influence the patient's arm, as heretofore. I made the attempt and failed. Dr. F—— then intimated a desire to have me put a question. I did so, as follows:

"M. Valdemar, can you explain to us what are your feelings or wishes now?"

There was an instant return of the hectic circles on the cheeks; the tongue quivered, or rather rolled violently in the mouth (although the jaws and lips remained rigid as before;) and at length the same hideous voice which I have already described, broke forth:

"For God's sake!—quick!—quick!—put me to sleep—or, quick!—waken me!—quick!—I say to you that I am dead!"

I was thoroughly unnerved, and for an instant remained undecided what to do. At first I made an endeavour to re-compose the patient; but, failing in this through total abeyance of the will, I retraced my steps and as earnestly struggled to awaken him. In this attempt I soon saw that I should be successful—or at least I soon fancied that my success would be complete—and I am sure that all in the room were prepared to see the patient awaken.

For what really occurred, however, it is quite impossible that any human being could have been prepared.

As I rapidly made the mesmeric passes, amid ejaculations of "dead! dead!" absolutely bursting from the tongue and not from the lips of the sufferer, his whole frame at once—within the space of a single minute, or even less, shrunk—crumbled—absolutely rotted away beneath my hands. Upon the bed, before that whole company, there lay a nearly liquid mass of loathsome—of detestable putrescence.

Edgar Allan Poe (1809–1849) helped pioneer the short story form, and was a trailblazer in the detective, horror, and science fiction genres. Hugo Gernsback, the founder of the modern science fiction magazine, formed an influential definition of the field in the first ever issue of "Amazing" (April 1926): "By 'scientifiction' I mean the Jules Verne, H. G. Wells, Edgar Allan Poe type of story". Poe's name continues to keep good company.

Commentary

In Chap. 10 we considered suspended animation. As we discussed there, several organisations now provide a cryonics service. The service provider, when presented with the dead body of a client, freezes the corpse. (Or, for those on a budget, just the head.) Why do people pay for this service? Well, they hope this frozen state will preserve their 'human-ness', that the onward march of science will one day permit reanimation, and that medical advances will provide a cure for whatever killed them. The idea of using cryonics to cheat death thus involves much wishful thinking and depends on technology that might never succeed. Critics, meanwhile, worry about the ethical issues involved and perhaps would prefer it if the whole approach led to a dead end. (Pun intended.)

Whether one supports or objects to cryonics, medics already practice a form of suspended animation. They just use it in specific cases and for short periods of time. Consider, for example, the technique called Emergency Preservation and Resuscitation (EPR), under trial by doctors in America to resuscitate people who suffer cardiac arrest from trauma. Under EPR, doctors drop the body temperature of a patient at risk of bleeding out, thereby slowing metabolic processes. This gives surgeons more time to operate. Even if the dreams of cryonics never comes to pass, procedures such as EPR will become more sophisticated. As medical science advances, doctors will have new possibilities for extending life. And this will lead to profound questions about the meaning of human existence.

In "The facts in the case of M. Valdemar" Poe invoked mesmerism, a form of hypnotism (see Fig. 11.1), to create an impression of verisimilitude. In 1845, when this story appeared, many considered mesmerism a form of reliable technology. Well, mention of mesmerism had the desired effect. After the story's publication, readers bombarded Poe with correspondence. They asked the question: Valdemar—fictional character or medical patient? Mention of mesmerism sounds quaint to a modern reader, but substitute modern technology for mesmerism and we see how Poe gave us a thought experiment with which to explore medical ethics. What interventions should doctors make as a person approaches death? Who should decide upon those interventions? If a person's body remains intact but they exhibit no signs of consciousness, do they still live? In short, how should we define death?

Sarah O'Dell, a physician–scholar who has studied for both PhD and MD degrees, explores the intersection of humanities with the world of ever-improving medical technology. In a lucid and thought-provoking essay she

Fig. 11.1 The French–British artist Alphonse Legros produced this etching in 1861, sixteen years after the initial publication of Poe's story, to illustrate Baudelaire's translation of the tale into French. Legros depicts the mesmerist, arms outstretched, placing the dying Valdemar into a hypnotic trance. The etching is without embellishment, and evokes the matter-of-fact approach of the story's narrator. Nowadays, of course, we give no credence to mesmerism. But a modern version of this etching, set in a high-tech hospital, might depict doctors hooking Valdemar up to various machines capable of keeping his heart pumping, his lungs breathing, and his body fed and watered. (Public domain)

discusses the question of 'undead bodies'. What if people disagree about whether a person has died? If one doctor declares a patient dead and another states the patient still lives, how should we think of that person? As alive? Dead? Undead? Her essay has relevance for us here because it draws comparisons between a case study written by the pulmonologist John M. Luce titled "The uncommon case of Jahi McMath" and Poe's story "The facts in the case of M. Valdemar".

Below, I provide a brief summary of the details of the Jahi McMath case.

Jahi, a 13-year-old Californian girl, suffered from obstructive sleep apnea. On 9 December 2013 she underwent complex surgery at Children's Hospital Oakland. Doctors hoped the procedures would improve airflow when she slept. After the surgery, Jahi began bleeding from the nose and mouth. The blood loss triggered a cardiac arrest, and she suffered a disruption of blood flow to the brain. She never regained consciousness. Three days after the surgery, doctors determined her to be brain dead. Doctors told her family that, since Jahi was legally dead, they would switch off the life-support systems that kept her heart beating and her lungs breathing. Her parents disputed the determination and argued their daughter still lived. A California court ordered an assessment by a doctor from a different hospital. On 23 December, an outside expert confirmed that the girl showed no cerebral activity. She could not breathe unaided. Jahi had died.

On 3 January 2014, the coroner's office issued a death certificate and gave the date of death as 12 December 2013. In the absence of an autopsy, the certificate carried no cause of death.

Jahi's parents went to higher courts in California to appeal the decision. They said Jahi remained alive because her heart still beat. They asked for the hospital to continue to provide life-support measures until the family could make other arrangements. In response, the hospital deemed it unethical to force doctors to provide medical care to a dead body. Following mediation, doctors released Jahi from the hospital into her mother's care. The hospital agreed that Jahi could stay on a ventilator and her intravenous fluid lines could remain. But they would not perform a tracheostomy or insert feeding tubes "on a corpse".

Jahi was dead in California. But not in New Jersey.

On the east and west coasts of the United States, the law differs. In New Jersey, doctors can only declare death on the grounds of irreversible cardiopulmonary activity. After Jahi left Children's Hospital Oakland, her parents relocated to the eastern state. There, surgeons performed the operation that Californian doctors had refused. Jahi underwent a tracheostomy, which permitted permanent attachment to a mechanical ventilator. A gastronomy tube allowed artificial nutrition.

For four and a half years, Jahi received almost round-the-clock medical care. Her mother applied skin care products and gave her various supplements. The family released more than 50 videos that appeared to show Jahi responding to requests. Jahi's mother would say "move your hand" or "kick your foot" and Jahi seemed to react appropriately. The family argued this proved Jahi lived. The medical establishment countered that Jahi displayed no volition. Those movements? Spinal cord reflexes.

After 4 years, Jahi's condition changed. Following exploratory surgery, her liver and kidneys failed. She suffered internal bleeding. Doctors ceased life support and her heart stopped. Authorities then issued a second death certificate, giving the date of death as 22 June 2018.

As the McMath case demonstrates, brain-dead patients on life-support machines might look alive. Their chests rise and fall as if breathing. Their skin feels warm to the touch. Bodily functions not dependent upon the brain, such as growth and sexual maturation, continue. The overwhelming majority of medical organisations, though, insist such a patient has zero chance of recovering consciousness. Without those life-support machines, those patients will not breathe by themselves. They have died.

Not all jurisdictions accept the medical consensus.

Some religions reject the notion of brain death, while different states and different countries operate under different definitions of what death means. That lack of harmony, as we have seen, can lead to problems. And the accelerating advance of medical science will make those problems more acute.

Surgeons have learned how to transplant the heart, intestines, kidney, liver, lung, pancreas, thymus and uterus. They have transplanted bones, corneas, nerves, skin, tendons, and veins. With each passing year, prosthetic devices improve in functionality and cosmetic appearance. Brain implants already help paralysed people to walk, amputees to control prosthetic arms, and those suffering from Parkinson's disease to moderate their symptoms. With each year, implants become smaller and more flexible (see Fig. 11.2). In the future

Fig. 11.2 In January 2024, a multidisciplinary team of researchers at University of California San Diego published details of a new neural implant, which consists of a flexible polymer strip packed with graphene electrodes. When placed on the surface of the brain, the implant can provide data about neural activity taking place deep inside the brain. Future neural implants will become smaller and minimally invasive, while providing detailed information about the brain to which they are attached. (Credit: David Baillot/UC San Diego Jacobs School of Engineering, CC BY-SA 4.0)

we can imagine neural implants the size of grains of dust, flooding a person's brain. Those implants could control a person's medical devices, certainly. They could keep an artificial heart beating, an artificial pancreas monitoring insulin levels, an artificial kidney filtering blood. Might they also record a person's thoughts, memories, and emotions? And perhaps synchronise those recordings with an AI system?

In a world of such technology, how should we define death?

Notes and Further Reading

readers bombarded Poe with correspondence—For a biography of Poe, see Ingram (1880).

drop the body temperature of a patient at risk—Emergency Preservation and Resuscitation (EPR) is a medical procedure that doctors are trialling to save the lives of people with who have suffered severe traumatic injury. An EPR patient might have suffered a gunshot wound, for example, and gone into cardiac arrest after heavy bleeding. In EPR, doctors reduce the patient's body temperature to as low 10 °C and replace the blood with ice-cold saline. At this point the patient has no pulse and no measurable brain activity. Surgeons then have hours rather than minutes to try and repair the trauma damage. EPR is not without its drawbacks, however, and the technique has been criticised on ethical grounds. See Tisherman (2022).

Sarah O'Dell, a physician–scholar—For a fascinating discussion of Poe's story in the context of developments in medical technology, see O'Dell (2020). For illustration, O'Dell draws on the Jahi McMath case; see Luce (2015).

details of the Jahi McMath case—Aviv (2018) presents a thorough account of the Jahi McMath case, and situates it within the broader debate about the meaning of death.

help paralysed people to walk—Researchers have demonstrated how the introduction of a 'digital bridge' between brain and spinal cord can help those with a spinal cord injury to walk again. See Lorach et al. (2023). The development of better, faster, smaller neural implants is taking place in university departments and research centres. Commercial companies are also interested in brain–computer interfaces. Perhaps the best known company operating in this space is Elon Musk's Neuralink; see Neuralink (n.d.).

References

Aviv, R.: What does it mean to die? The New Yorker. (Appears in the February print issue under the title *The death debate*) www.newyorker.com/magazine/2018/02/05/what-does-it-mean-to-die (2018)

Ingram, J.H.: Edgar Allan Poe: His Life, Letters, and Opinions. John Hogg Kennedy, London (1880)

Lorach, H., et al.: Walking naturally after spinal cord injury using a brain–spine interface. Nature. **618**, 126–133 (2023)

Luce, J.M.: The uncommon case of Jahi McMath. Chest: Medical Ethics. **147**(4), 1144–1151 (2015)

Neuralink: Homepage. https://neuralink.com (n.d.)

O'Dell, S.: "The facts in the case of M. Valdemar": undead bodies and medical technology. J. Med. Humanities. **41**, 229–242 (2020). https://doi.org/10.1007/s10912-019-09574-w

Tisherman, S.A.: Emergency preservation and resuscitation for cardiac arrest from trauma. Ann. N. Y. Acad. Sci. **1509**(1), 5–11 (2022). https://doi.org/10.1111/nyas.14725

12

Humans Supplanted

The Horla (Guy de Maupassant)

MAY 8. What a glorious day! I have spent all the morning lying on the grass in front of my house, under the enormous plantain tree which covers and shades and shelters the whole of it. I like this part of the country, and I am fond of living here because I am attached to it by deep roots, those profound and delicate roots that anchor a man to the soil on which his ancestors were born and died, to their traditions, their usages, their food, the local expressions, the peculiar intonations of the peasants, the smell of the soil, the hamlets, and the very air itself.

I love the house in which I grew up. From my windows I can see the Seine, which flows by the side of my garden, on the other side of the road, almost through my grounds, the great and wide Seine, which goes to Rouen and Havre, and which is covered with boats passing to and fro.

On the left, down yonder, lies Rouen, populous Rouen with its blue roofs massing under pointed, Gothic towers. They are beyond number, delicate or sturdy, dominated by the spire of the cathedral, full of bells which sound through the blue air on fine mornings, sending their sweet and distant iron clang to me, their metallic sounds, now stronger and now weaker, according as the wind is strong or light.

What a beautiful morning it has been!

About eleven o'clock, a long line of boats drawn by a steam-tug that looked like a fly, scarcely puffing while it belched thick clouds of smoke, passed my gate.

After two English schooners, whose red flags fluttered toward the sky, there came a magnificent Brazilian three-master, all white and wonderfully clean and shining. I saluted it, I hardly know why, except that the sight of the vessel filled me with joy.

MAY 12. I have had a slight fever for the last few days, and I feel ill, or rather I feel low-spirited.

Whence come those mysterious influences which change our happiness into discouragement, and our self-confidence into diffidence? One might almost say that the air, the invisible air, is full of unknowable forces, whose mysterious presence we have to endure. I wake up in the best of spirits, with an inclination to sing in my heart. Why? I go down by the side of the water, and suddenly, after walking a short distance, I return home wretched, as if some misfortune were awaiting me there. Why? Is it a cold shiver which, passing over my skin, has upset my nerves and given me a fit of low spirits? Is it the form of the clouds, or the tints of the sky, or the colours of the surrounding objects which are so changeable, which have troubled my thoughts as they passed before my eyes? Who can tell? Everything that surrounds us, everything that we see without looking at it, everything that we touch without knowing it, everything that we handle without feeling it, everything that we meet without clearly distinguishing it, has a rapid, surprising, and inexplicable effect upon us and upon our organs, and through them on our ideas and on our being itself.

How profound it is, that mystery of the invisible! We cannot fathom it with our miserable senses. Our eyes are unable to perceive what is either too small or too great, too near or too far from us; we can see neither the inhabitants of a star nor of a drop of water. Our ears deceive us, for they transmit vibrations of the air to us as sonorous notes. Our senses are fairies who work the miracle of changing that movement into noise, and by that metamorphosis give birth to music, which makes the mute agitation of nature a harmony. So with our sense of smell, which is weaker than that of a dog, and so with our sense of taste, which can scarcely distinguish the age of a wine!

If only we had other organs which could work other miracles in our favour, what a number of fresh things we might discover around us!

MAY 16. I am ill, decidedly! I was so well last month! I am feverish, horribly feverish, or rather I am in a state of feverish enervation, which makes my mind suffer as much as my body. I have without ceasing the horrible sensation of some danger threatening me, the apprehension of some coming misfortune or of approaching death, a presentiment which is no doubt an attack of some illness still unnamed, which germinates in the flesh and in the blood.

12 Humans Supplanted

MAY 18. I have just come from consulting my doctor, for I can no longer sleep. He found that my pulse was high, my eyes dilated, my nerves highly strung, but no alarming symptoms. I must have a course of shower baths and of bromide of potassium.

MAY 25. No change! My state is really very peculiar. As night falls, an incomprehensible feeling of unease seizes me, as if the night concealed some terrible menace toward me. I dine quickly, and then try to read, but I do not understand the words, and can scarcely distinguish the letters. Then I walk up and down my drawing-room, oppressed by a feeling of confused and irresistible fear, a fear of sleep and a fear of my bed.

About ten o'clock I go up to my room. As soon as I have entered I lock and bolt the door. I am frightened—of what? I have never been frightened of anything before now. I open my cupboards, and look under my bed; I listen—I listen—to what? How strange it is that a simple feeling of discomfort, of impeded or heightened circulation, perhaps the irritation of a nervous centre, a slight congestion, a small disturbance in the imperfect and delicate functions of our living machinery, can turn the most light-hearted of men into a melancholy one, and make a coward of the bravest? Then, I go to bed, and I wait for sleep as a man might wait for the executioner. I wait for its coming with dread, and my heart beats and my legs tremble, while my whole body shivers beneath the warmth of the bedclothes, until the moment when I suddenly fall asleep, as a man throws himself into a pool of stagnant water in order to drown. I do not feel this perfidious sleep coming over me as I used to, but a sleep which is close to me and watching me, which is going to seize me by the head, to close my eyes and annihilate me.

I sleep—a long time—two or three hours perhaps—then a dream—no—a nightmare lays hold on me. I feel that I am in bed and asleep—I feel it and I know it—and I feel also that somebody is coming close to me, is looking at me, touching me, is getting on to my bed, is kneeling on my chest, is taking my neck between his hands and squeezing it—squeezing it with all his might in order to strangle me.

I struggle, bound by that terrible powerlessness which paralyses us in our dreams. I try to cry out—but I cannot; I want to move—I cannot. I try, with the most violent efforts and out of breath, to turn over and throw off this being which is crushing and suffocating me—I cannot!

And then suddenly I wake up, shaken and bathed in perspiration. I light a candle and find that I am alone. And after that crisis, which occurs every night, I eventually fall asleep and slumber tranquilly till morning.

JUNE 2. My state has grown worse. What is the matter with me? The bromide does me no good, and the shower-baths have no effect whatever.

Sometimes, in order to tire myself out, though I am fatigued enough already, I go for a walk in the forest of Roumare. I used to think at first that the fresh light and soft air, impregnated with the door of herbs and leaves, would instil new life into my veins and impart fresh energy to my heart. One day I turned into a broad ride in the wood, and then I diverged toward La Bouille, through a narrow path, between two rows of exceedingly tall trees, which placed a thick, green, almost black roof between the sky and me.

A sudden shiver ran through me, not a cold shiver, but a shiver of agony, and so I hastened my steps, uneasy at being alone in the wood, frightened stupidly and without reason, at the profound solitude. Suddenly it seemed as if I were being followed, that somebody was walking at my heels, close, quite close to me, near enough to touch me.

I turned round suddenly, but I was alone. I saw nothing behind me except the straight, broad ride, empty and bordered by high trees, horribly empty. On the other side also it extended until it was lost in the distance, and looked just the same—terrible.

I closed my eyes. Why? And then I began to turn round on one heel very quickly, just like a top. I nearly fell down, and opened my eyes. The trees were dancing round me and the earth heaved. I was obliged to sit down. Then, ah! I no longer remembered how I had come! What a strange idea! What a strange, strange idea! I did not the least know. I started off to the right, and got back into the avenue which had led me into the middle of the forest.

JUNE 3. I have had a terrible night. I shall go away for a few weeks, for no doubt a journey will put me right.

JULY 2. I have come back, quite cured, and have had a most delightful trip into the bargain. I have been to Mont Saint-Michel, which I had not seen before.

What a sight, when one arrives as I did, at Avranches toward the end of the day! The town stands on a hill, and I was taken into the public garden at the extremity of the town. I uttered a cry of astonishment. An extraordinarily large bay lay extended before me, as far as my eyes could reach, between two hills which were lost to sight in the mist; and in the middle of this immense yellow bay, under a clear, golden sky, a peculiar hill rose up, somber and pointed in the midst of the sand. The sun had just disappeared, and under the still flaming sky stood out the outline of that fantastic rock which bears on its summit a picturesque monument.

At daybreak I went to it. The tide was low, as it had been the night before, and I saw that wonderful abbey rise up before me as I approached it. After several hours' walking, I reached the enormous mass of rock which supports the little town, dominated by the great church. Having climbed the steep and narrow street, I entered the most wonderful Gothic building that has ever been erected to God on earth, large as a town, and full of low rooms which seem buried beneath vaulted roofs, and of lofty galleries supported by delicate columns.

I entered this gigantic granite jewel, which is as light in its effect as a bit of lace and is covered with towers, with slender belfries to which spiral staircases ascend. The flying buttresses raise strange heads that bristle with chimeras, with devils, with fantastic animals, with monstrous flowers, are joined together by finely carved arches, to the blue sky by day, and to the black sky by night.

When I had reached the summit, I said to the monk who accompanied me: "Father, how happy you must be here!" And he replied: "We get strong winds, Monsieur". And so we began to talk while watching the rising tide, which ran over the sand and covered it with a steel cuirass.

And then the monk told me stories, all the old stories belonging to the place—legends, nothing but legends.

One of them struck me forcibly. The country people, those belonging to the Mornet, declare that at night one can hear talking going on in the sand, and also that two goats bleat, one with a strong, the other with a weak voice. Incredulous people declare that it is nothing but the screaming of the sea birds, which occasionally resembles bleatings, and occasionally human lamentations. But benighted fishermen swear they have met an old shepherd, whose cloak covered head they can never see, wandering on the sand, between two tides, round the little town placed so far out of the world. They declare he is guiding and walking before a he-goat with a man's face and a she-goat with a woman's face, both with white hair, who talk incessantly, quarrelling in a strange language, and then suddenly cease talking in order to bleat with all their might.

"Do you believe it?" I asked the monk.

"I scarcely know," he replied.

I continued: "If there are other beings besides ourselves on this earth, how comes it that we have not known it for so long a time, or why have you not seen them? How is it that I have not seen them?"

He replied: "Do we see the hundred-thousandth part of what exists? Look here; there is the wind, which is the strongest force in nature. It knocks down men, and blows down buildings, uproots trees, raises the sea into mountains of water, destroys cliffs and casts great ships on to the breakers; it kills, it whistles, it sighs, it roars. But have you ever seen it, and can you see it? Yet it exists for all that."

I was silent before this simple reasoning. That man was a philosopher, or perhaps a fool. I could not say which exactly, so I held my tongue. What he had said had often been in my own thoughts.

JULY 3. I have slept badly. Certainly there is some feverish influence here, for my coachman is suffering in the same way as I am. When I went back home yesterday, I noticed his singular paleness, and I asked him: "What is the matter with you, Jean?"

"The matter is that I never get any rest, and my nights devour my days. Since your departure, Monsieur, there has been a spell over me."

The other servants are all well, but I am very frightened of having another attack, myself.

JULY 4. I am decidedly taken again. My old nightmares have returned. Last night I felt somebody leaning on me, sucking my life from between my lips with his mouth. Yes, he was sucking it out of my neck like a leech would have done. Then he got up, satiated. I woke up, so beaten, crushed, and annihilated that I could not move. If this continues for a few days, I shall certainly go away again.

JULY 5. Have I lost my reason? What has happened? What I saw last night is so strange that my head wanders when I think of it!

As I do now every evening, I had locked my door. Then, being thirsty, I drank half a glass of water. I happened to notice that the carafe was full up to the cut-glass stopper.

I went to bed and fell into one of my terrible sleeps, from which I was aroused in about two hours by a still more terrible shock.

Picture to yourself a sleeping man who is being murdered, who wakes up with a knife in his chest, a gurgling in his throat, is covered with blood, can no longer breathe, is going to die and does not understand anything at all about it—there you have it.

Having recovered my senses, I was thirsty again, so I lit a candle and went to the table where I had placed my carafe. I lifted up the carafe and tilted it over my glass, but nothing came out. It was empty! It was completely empty! At first I could not understand it at all. Suddenly I was seized by such a terrible feeling that I had to sit down, or rather fall into a chair! I sprang up with a bound to look about me. Then I sat down again, overcome by astonishment and fear, in front of the transparent crystal carafe! I looked at it with fixed eyes, trying to solve the puzzle. My hands trembled! Somebody had drunk the water. But who? I? It must have been me. Who else could it have been? In that case I was a somnambulist—was living, without knowing it, that double, mysterious life which makes us doubt whether there are not two beings in us—whether a strange, unknowable, and invisible being does not, during our moments of mental and physical torpor, animate the inert body, forcing it to a more willing obedience than it yields to ourselves.

Oh, can anyone understand my horrible agony? Can anyone understand the emotion of a man sound in mind, educated, rational, staring in terror through the glass of his carafe at the disappearance of a little water while he slept? I remained sitting until daylight, not daring to go to bed again.

JULY 6. I am going mad. My carafe was emptied again last night—or rather, I emptied it.

But is it I? Is it I? Who could it be? Who? Oh, God! Am I going mad? Who will save me?

JULY 10. I have just been through some surprising ordeals. Undoubtedly I must be mad! And yet ...

On July 6, before going to bed, I put some wine, milk, water, bread, and strawberries on my table. Somebody drank—I drank—all the water and a little of the milk. But neither the wine, nor the bread, nor the strawberries were touched.

On July 7, I renewed the same experiment and got the same results. On July 8, I left out the water and the milk. Nothing was touched.

Lastly, on July 9, I put only water and milk on my table, taking care to wrap up the bottles in white muslin and to tie down the stoppers. Then I rubbed my lips, my beard, and my hands with pencil lead, and went to bed.

Deep slumber seized me, soon followed by a terrible awakening. I had not moved, and my sheets were not marked. I rushed to the table. The muslin round the bottles remained intact. I undid the string, trembling with fear. All the water had been drunk, and so had the milk! Oh, my God!

I am leaving for Paris immediately.

JULY 12. Paris. I must have lost my head during the last few days. I must be the plaything of my enervated imagination, unless I am really a somnambulist, or I have been brought under the power of one of those influences—hypnotic suggestion, for example—which are known to exist, but have hitherto been inexplicable. In any case, my mental state bordered on madness, but twenty-four hours of Paris sufficed to restore me to my equilibrium.

Yesterday, after doing some business and making various calls, which instilled fresh and invigorating mental air into me, I wound up my evening at the Théâtre Français. A drama by Alexander Dumas the Younger was being acted, and his brilliant and powerful play completed my cure. Without doubt, loneliness is dangerous to active minds. We need around us men who can think and can talk. When we live alone for a long time, we people the void with phantoms.

I walked along the boulevards back to my hotel in excellent spirits. Amid the jostling of the crowd I thought, not without irony, of my terrors and surmises of the previous week, when I believed, yes, I believed, that an invisible being lived beneath my roof. How weak our mind is. How quickly it is terrified and unbalanced as soon as we are confronted with a small, incomprehensible fact. Instead of dismissing the problem with the simple words: "We do not understand because we cannot find the cause," we immediately imagine terrible mysteries and supernatural powers.

JULY 14. Fête de la Republic. I walked through the streets, and the crackers and flags amused me like a child. Still, it is very foolish to make merry on a set date, by government decree. People are like a flock of sheep, now steadily patient, now in ferocious revolt. Say to it: "Amuse yourself," and it amuses itself. Say to it: "Go and fight with your neighbour," and it goes and fights. Tell it: "Vote for the Emperor," and it votes for the Emperor. Then tell it: "Vote for the Republic," and it votes for the Republic.

Those who rule are stupid, too; but instead of obeying men they obey principles, a course which can only be foolish, ineffective, and false, for the very reason that principles are ideas which are considered as certain and unchangeable, whereas in this world one is certain of nothing, since light is an illusion and noise is deception.

JULY 16. I saw some things yesterday that troubled me greatly.

I was dining at my cousin's, Madame Sablé, whose husband is colonel of the Seventy-sixth Chasseurs at Limoges. I met two young women there, one of whom had married a medical man, Dr. Parent, who devotes himself largely to nervous diseases and to the extraordinary manifestations which just now experiments in hypnotism and suggestion are producing.

He told us at some length about the amazing results obtained by English scientists and the doctors of the medical school at Nancy, and the facts which he adduced appeared to me so fantastic that I declared that I was altogether incredulous.

"We are," he declared, "on the point of discovering one of the most important secrets of nature, I mean to say, one of its most important secrets on this earth, for assuredly there are some up in the stars, yonder, of a different kind of importance. Ever since man has thought, since he has been able to express and write down his thoughts, he has felt himself close to a mystery which is impenetrable to his coarse and imperfect senses, and he endeavours to supplement the feeble penetration of his organs by the efforts of his intellect. As long as that intellect remained in its elementary stage, this intercourse with invisible spirits assumed forms which were commonplace though terrifying. Thence sprang the popular belief in the supernatural, the legends of wandering spirits, of fairies, of gnomes, of ghosts, I might even say the conception of God, for our ideas of the artificer–creator, from whatever religion they may have come down to us, are certainly the most mediocre, the stupidest, and the most unacceptable inventions that ever sprang from the frightened brain of any human creature. Nothing is truer than Voltaire's saying: 'If God made man in His own image, man has certainly paid Him back in kind.'

"But for rather more than a century, men seem to have had a presentiment of something new. Mesmer and some others have put us on a fresh path, and within the last few years in particular we have arrived at some surprising results."

My cousin, as incredulous as I, smiled. Dr. Parent said to her: "Would you like me to try and send you to sleep, Madame?"

"Yes, do."

She sat down in an armchair, and he began to look at her fixedly, as if to fascinate her. I suddenly felt myself somewhat discomposed: my heart began to beat rapidly, my throat constricted. I saw that Madame Sablé's eyes were growing heavy, her mouth twitched, and her bosom heaved. Within ten minutes she was asleep.

"Go behind her," the doctor said to me, so I took a seat behind her. He put a visiting-card into her hands, and said to her: "This is a looking-glass. What do you see in it?"

She replied: "I see my cousin."

"What is he doing?"

"He is twisting his moustache."

"And now?"

"He is taking a photograph out of his pocket."

"Whose photograph is it?"

"His own."

That was true, for the photograph had been given me that same evening at the hotel.

"What is his attitude in this portrait?"

"He is standing, with his hat in his hand."

She saw these things in that card, in that piece of white pasteboard, as if she had seen them in a looking-glass.

The young women, terrified, cried: "That is quite enough! Quite, quite enough!"

But the doctor said to her authoritatively: "You will get up at eight o'clock tomorrow morning. Then you will go and call on your cousin at his hotel and ask him to lend you the five thousand francs which your husband asks of you, and which he will ask for when he sets out on his coming journey."

Then he woke her up.

On returning to my hotel, I thought over this curious séance and I was assailed by doubts, not as to my cousin's absolute and undoubted good faith, for I had known her as well as if she had been my own sister ever since she was a child, but as to a possible trick on the doctor's part. Had he perhaps kept a glass hidden in his hand, which he showed to the young woman in her sleep at the same time as he did the card? Professional conjurers do things as strange.

I went back to the hotel and to bed. This morning, at about half past eight, I was awakened by my footman, who said to me: "Madame Sablé has asked to see you immediately, Monsieur." I dressed hastily and went to her.

She sat down in some agitation, with her eyes on the floor, and without raising her veil said to me: "My dear cousin, I am going to ask a great favour of you."

"What is it, my dear?"

"I do not like to tell you, and yet I must. I am in absolute want of five thousand francs."

"What, you?"

"Yes, I, or rather my husband, who has asked me to procure them for him."

I was so stupefied that I hesitated to answer. I asked myself whether she had not really been making fun of me with Dr. Parent, if it were not merely a very well-acted farce which had been got up beforehand. On looking at her attentively, however, my doubts disappeared. She was trembling with grief, so painful was this step to her, and I saw that her throat was quivering with sobs.

I knew she was rich, so I continued: "What! Has your husband not got five thousand francs at his disposal? Come, think. Are you sure he commissioned you to ask me for them?"

She hesitated for a few moments, as if she were making a great effort to search her memory, and then she replied: "Yes—yes, I am quite sure of it."

"He has written to you?"

She hesitated again, reflecting. I guessed the torture of her thoughts. She did not know. She only knew that she was to borrow five thousand francs of me for her husband. She plucked up the courage to lie.

"Yes, he has written to me."

"When, pray? You did not mention it to me yesterday."

"I got his letter this morning."

"Can you show it to me?"

"No. No … no … it contained private matters, things too personal to ourselves. I burned it."

"So your husband runs into debt?"

She hesitated again, then murmured: "I don't know."

I said abruptly: "I can't lay my hands on five thousand francs at this moment, my dear."

An agonised wail broke from her. She said: "Oh, I implore you, I implore you to get the money for me."

She grew dreadfully excited, clasping her hands as if she were praying to me. The tone of her voice changed: she wept, stammering, torn with grief, goaded by the irresistible order that had been laid on her.

"Oh, I beg you to get it … if you only knew what I am suffering … I must have the money today."

I took pity on her: "You shall have the money at once, I swear to you."

"Oh, thank you! Thank you! How kind you are."

"Do you remember," I went on, "what took place at your house last night?"

"Yes."

"Do you remember that Dr. Parent sent you to sleep?"

"Yes."

"Oh! Very well then. He ordered you to come to me this morning to borrow five thousand francs. You are now obeying that suggestion."

She considered this for a moment, then replied: "But it is my husband who wants the money."

For a whole hour I tried to convince her, but could not succeed. When she had gone I went to the doctor. He was just going out, and he listened to me with a smile. He said: "Now do you believe?"

"Yes, I must."

"Let's pay a call on your cousin."

She was already resting on a couch, overcome with fatigue. The doctor felt her pulse, and looked at her for some time with one hand raised toward her eyes. By degrees her eyes closed under the irresistible power of this magnetic influence. When she was asleep, he said: "Your husband no longer require five thousand francs! You must, therefore, forget that you asked your cousin to lend them to you. If your cousin speaks to you about it, you will not understand him."

Then he woke her up. I took out a notecase and said: "Here is what you asked me for this morning, my dear." But she was so surprised, that I did not venture to persist. I did, however, try to rouse her memory, but she denied it fiercely, thought I was making fun of her, and in the end, very nearly lost her temper.

Back at the hotel. This experiment disturbed me so much I could not bring myself to take lunch.

JULY 19. I told several people about the adventure and they laughed at me. I no longer know what to think. The wise man says: Perhaps?

JULY 21. I dined at Bougival, then spent the evening at a boatmen's ball. Decidedly everything depends on place and surroundings. It would be the height of folly to believe in the supernatural on the Ile de la Grenouillière. But on the top of Mont Saint-Michel? Or in India? We are terrified under the influence of our surroundings. I shall return home next week.

JULY 30. I returned to my own house yesterday. All is well.

AUGUST 2. Nothing fresh. The weather is splendid. I spend my days watching the Seine flow past.

AUGUST 4. The servants are quarrels amongst themselves. They declare that someone breaks the glasses in the cupboards at night. The footman accuses the cook, she accuses the needlewoman, and the latter accuses the other two. Who is the culprit? It would take a clever person to tell.

AUGUST 6. This time, I am not mad. I have seen—I have seen—I have seen *something*! I can doubt no longer—I have seen it!

I was walking in my rose garden, between the autumn roses which are just beginning to come out. It was two o'clock, in the full sunlight. As I stopped to look at a *Géant de Bataille*, which had three splendid blooms, I distinctly saw the stalk of one of the roses bend close to me, as if an invisible hand had twisted it, and then it broke, as if that hand had plucked it! Then the flower rose, describing in the air the curve that a hand would have made carrying it toward a mouth. It remained suspended in the clear air, alone, motionless, a terrifying scarlet splash, three yards from my eyes. In desperation I rushed at it, grasping at it. My fingers closed on nothing. It had disappeared. A savage rage seized me, a rage against myself. A rational, sober man does not suffer such hallucinations.

But was it an hallucination? I turned to look for the flower, and I found it immediately under the bush, freshly broken, between the two other roses that remained on the branch. I went back to the house with a much disturbed mind. I am now certain, certain as I am that night follows day, that there exists close to me an invisible being who lives on milk and on water, who can touch objects, take them and move them from one place to another. A being who is, consequently, endowed with a material nature, although imperceptible to sense, and who lives as I do, under my roof …

AUGUST 7. I slept tranquilly. He drank the water out of my carafe, but did not disturb my sleep.

I ask myself whether I am mad. As I was walking just now in the sun, by the riverside, I doubted my sanity. These were not vague doubts I have had before, but precise and absolute doubts. I have seen mad people, and I know some who were intelligent, lucid, even clear-sighted in every concern of life, except on one point. They could speak clearly, readily, profoundly on everything, until their thoughts were caught in the breakers of their delusions and there went to pieces, were dispersed and swamped in that furious and terrible sea of fogs and squalls which is called madness.

I should certainly think that I was mad, absolutely mad, were not conscious that I knew my state, if I could not fathom it and analyse it with the most complete lucidity. I must be, in fact, a reasonable man labouring under an hallucination. Some unknown disturbance must have been excited in my brain, one of those disturbances that physiologists of the present day try to note and to fix precisely, and the disturbance must have caused a profound gulf in my mind and in the orderly and logical working of my thoughts. Similar phenomena occur in dreams, and lead us through the most unlikely phantasmagoria, without causing us any surprise, because our mechanism of judgement, the controlling censor, is asleep while our imaginative faculty wakes and works. Could one of the invisible strings that control my mental keyboard have become muted?

Some men, after an accident, lose the recollection of proper names, or of verbs, or of numbers, or merely of dates. The localization of all the avenues of thought is now proved. Is there anything surprising, therefore, in the idea that my faculty of examining the unreality of certain hallucinations has stopped working in my brain just now?

I thought of all this as I walked by the side of the water. The sun shone brightly on the river and clothed the earth with beauty, while it filled me with love of life, of the swallows whose swift flight always delights my eyes, of the riverside grasses whose rustling is a pleasure to my ears.

By degrees, however, an inexplicable feeling of unease seized me. It felt as if some unknown force were numbing me, stopping me, preventing me from going further, calling me back. I felt the painful wish to return that comes on when you have left a beloved invalid at home, and are seized by a presentiment that the illness has taken a turn for the worse.

So I returned, despite myself, feeling certain I should find some bad news awaiting me, a letter or a telegram. There was nothing. And I was left more surprised and uneasy than if I had had another fantastic vision.

AUGUST 8. Yesterday I spent a terrible evening. He did not show himself again, but I feel He is near me, watching me, spying on me, taking possession of me, dominating me, and more to be feared when He hides himself this way than if He were to manifest his constant and invisible presence by supernatural phenomena.

However, I slept.

AUGUST 9. Nothing, but I am afraid.

AUGUST 10. Nothing, but what will happen tomorrow?

AUGUST 11. Still nothing. I cannot stop at home with this fear hanging over me and these thoughts in my mind. I shall go away.

AUGUST 12. Ten o'clock at night. All day long I have been trying to get away, and have not been able. I contemplated a simple and easy act of liberty, a carriage ride to Rouen—and I was unable to do it. Why?

AUGUST 13. When one is attacked by certain maladies, the springs of our physical being seem broken, our energies destroyed, our muscles relaxed, our bones as soft as our flesh, and our blood as liquid as water. In some strange and distressing manner I am experiencing the same in my moral being. I no longer have any strength, any courage, any self-control, no power even summon up my will. I can't will anything. But someone wills for me—and I obey.

AUGUST 14. I am lost! Somebody possesses my soul and governs it! Somebody orders all my acts, all my movements, all my thoughts. I am no longer master of myself, nothing except an enslaved and terrified spectator of the things which I do. I wish to go out; I cannot. He does not wish to and so I remain, trembling and distracted in the armchair in which he keeps me sitting. I merely wish to get up and to rouse myself, so as to think that I am still master of myself: I cannot! I am riveted to my chair, and my chair adheres to the floor in such a manner that no force of mine can move us.

Then suddenly, I must, I must go to the foot of my garden to pick some strawberries and eat them—and I go there. I pick the strawberries and I eat them! Oh, my God! My God! Is there a God? If there is one, deliver me, save me, help me! Pardon me! Pity me! Have mercy on me! How I suffer! How I am tortured! How terrible this is!

AUGUST 15. Certainly this is the way in which my poor cousin was possessed and swayed, when she came to borrow five thousand francs from me. She was under the power of a strange will that had entered into her, like another soul, a parasitic and ruling soul. Is the world coming to an end?

But who is he, this invisible being that rules me, this unknowable being, this wanderer from a supernatural race?

So invisible beings exist? How is it, then, that since the beginning of the world they have never manifested themselves in such unmistakeable a manner as they do to me? I have never read of anything that resembles what is happening in my house. If only I could leave it, if I could only go away, flee far away and never return, I should be saved. But I cannot.

AUGUST 16. Today I managed to escape for two hours, like a prisoner who finds the door of his dungeon accidentally open. I suddenly felt I was free and that He was far away, and so I gave orders to put the horses in as quickly as possible, and I drove to Rouen. Oh, how delightful to be able to say to my coachman: "Go to Rouen!" and be obeyed!

I made him pull up before the library, and there I begged them to lend me Dr. Herrmann Herestauss's treatise on the unknown inhabitants of the ancient and modern world.

Then, as I was getting into my carriage, I intended to say: "To the railway station!" but instead of this I shouted—I did not speak I shouted—in such a loud voice that all the passers-by turned round: "Home!" and I fell back onto the cushion of my carriage, overcome by mental agony. He had found me out and regained possession of me.

AUGUST 17. What a night! What a night! And yet it seems to me that I ought to rejoice. I read until one o'clock in the morning! Herestauss, Doctor of Philosophy and Theogony, wrote the history and the manifestation of all those invisible beings that hover around man, or of whom he dreams. He describes their origins, their domains, their power; but none of them resembles the one that haunts me. One might say that man, ever since he has thought, has had a foreboding and a fear of a new being, stronger than himself, his successor in this world, and that, feeling him near, and not being able to foretell the nature of the unseen one, he has, in his terror, created the whole race of hidden beings, vague phantoms born of fear.

Having, therefore, read until one o'clock in the morning, I went and sat down at the open window, in order to cool my forehead and my thoughts in the calm night air. It was fine and warm. In earlier times I should have loved such a night!

There was no moon. The stars glittered in the black depths of the sky. Who inhabits those worlds? What forms, what living beings, what animals dwell there? Do the philosophers on those distant worlds know more than we do? What can they do more than we? What do they see that we do not know of? Will not one of them, some day, cross the gulf of space and appear on our earth as a conqueror, just as formerly the Norsemen crossed the sea to subjugate nations feebler than themselves?

We are so weak, so powerless, so ignorant, so small—we who live on this particle of mud which revolves in liquid air.

I fell asleep, dreaming thus in the cool night air, and then, having slept for about three quarters of an hour, I opened my eyes without moving, awakened by an indescribably confused and strange sensation. At first I saw nothing, and then suddenly it appeared to me as if a page of the book lying open on my table turned over of its own accord. Not a breath of air had come in at my window. I was surprised, and I sat waiting. About four minutes later I saw, I saw—yes I saw with my own eyes—another page lift itself up and fall down on the others, as if a finger had turned it over. My armchair was empty, appeared empty, but I knew He was there, sitting in my place, reading. With a furious bound, the lunge of an enraged wild beast that wishes to disembowel its tamer, I crossed my room to seize him, to strangle him, to kill him! But before I could reach it, my chair fell over as if somebody had fled from me. My table rocked, my lamp fell and went out, and my window closed as if some thief had been surprised and had ran out into the night, shutting it behind him.

So He had run away. He had been afraid. He, afraid of me!

Tomorrow, then … or the day after … or some day … I should be able to hold him in my clutches and crush him against the ground! Do not dogs occasionally fly at their masters' throats?

AUGUST 18. I have been thinking things over all day. Oh! yes, I will obey Him, follow His impulses, fulfil all His wishes, make myself humble, submissive, a coward. He is the stronger. But an hour will come …

AUGUST 19. I know now, I know, I know everything! I have just read the following in the *Revue du Monde Scientifique*:

"A curious piece of news comes to us from Rio de Janeiro. Madness, an epidemic of madness, comparable to the contagious madness that attacked the people of Europe in the Middle Ages, is at this moment raging in the Province of San-Paulo. The frightened inhabitants are leaving their houses, deserting their villages, abandoning their land, saying that they are pursued, possessed, governed like human cattle by invisible but tangible beings, vampires of some kind, who feed on their vitality while they sleep, in addition to drinking milk and water without, apparently, touching any other nourishment.

"Professor Don Pedro Henriques, accompanied by several medical savants, has gone to the district of San-Paulo, to study in situ the origin and the manifestations of this surprising madness, and to propose such measures to the Emperor as may appear to him to be most fitted to restore the delirious population to reason."

Ah! I remember! I remember now the fine Brazilian three-master that passed in front of my windows as it was going up the Seine, on the eighth of last May! I thought it looked so pretty, so white and bright! The Being was on board her, coming from there, where its race sprang from. And He saw me! He saw my house, white like the ship, and He sprang from the vessel on to the land. Oh, my God!

Now I know, I understand. The reign of man is over. He is here, whom the dawning fears of primitive peoples taught them to dread. He who was exorcised by troubled priests, whom sorcerers evoked on dark nights without seeing Him appear, He to whom the imaginations of the transient masters of the world lent all the monstrous or graceful forms of gnomes, spirits, genii, fairies, and familiar spirits. After the coarse conceptions of primitive fear, men more enlightened gave him a truer form. Mesmer divined him, and ten years ago physicians accurately discovered the nature of his power, even before He exercised it himself. They played with that weapon of their new Lord, the sway of a mysterious will over the human soul, which had become enslaved. They called it mesmerism, hypnotism, suggestion, whatever you like. I have seen them amusing themselves like foolish children with this horrible power. Woe to us! Cursed is man! He is here, the—the—what does He call himself—the—I fancy he is shouting out his name to me and I do not hear him—the—yes—He is shouting it out—I am listening—I cannot—repeat—it—Horla—I have heard—the Horla—it is He—the Horla—He has come!

Ah the vulture has been used to eat the dove, the wolf to eat the lamb; the lion to devour the sharp-horned buffalo; man to kill the lion with an arrow, a spear, a gun. But the Horla will make of man what man has made of the horse and of the ox: His chattel, His slave, and His food, by the mere power of His will. Woe to us!

But sometimes the beast rebels and kills his tamer. I should also like—I could—but I must know Him, touch Him, see Him. Learned men say that eye of the beast is different to ours and do not see as ours do. And my eye cannot distinguish this newcomer who is oppressing me.

Why? Oh, now I remember the words of the monk at Mont Saint-Michel: "Do we see the hundred-thousandth part of what exists? Look here; there is the wind, which is the strongest force in nature. It knocks down men, and blows down buildings, uproots trees, raises the sea into mountains of water, destroys cliffs and casts great ships on to the breakers; it kills, it whistles, it sighs, it roars. But have you ever seen it, and can you see it? Yet it exists for all that".

And I considered further: my eyes are so weak, so imperfect, that they do not even distinguish hard bodies that have the transparency of glass. If a looking-glass without quicksilver behind it were to bar my way, I should run into it, just like a bird that has flown into a room and breaks its head against the windowpanes. A thousand things can deceive a man and lead him astray. Why then should it be surprising if he cannot perceive a new body that offers no resistance to the passage of light?

A new being! Why not? It was assuredly bound to come! Why should we be the last? Why do we not distinguish it, like all the others created before us? The reason is that its nature is more delicate, its body finer and more finished than ours. Our makeup is so weak, so awkwardly conceived; our body is encumbered with organs that are always tired, always being strained like locks that are too complicated; it lives like a plant and like an animal nourishing itself with difficulty on air, herbs, and flesh; it is a living machine subject to sickness, deformity, decay; it is broken-winded, badly regulated, simple and eccentric, ingeniously yet badly made, a coarse and yet a delicate mechanism, in brief, the outline of a being which might become intelligent and great.

There are only a few—so few—stages of development in this world, from the oyster up to man. Why should there not be one more, once that period is over which separates each successive appearance of a species, one from the other?

Why not one more? Why not, also, other trees with immense, splendid flowers, perfuming whole regions? Why not other elements beside fire, air, earth, and water? There are four, only four, sources of our being! What a pity! Why should not there be forty, four hundred, four thousand? How poor, niggardly and brutish is life!—grudgingly given, poorly conceived, clumsily executed! Consider the grace of the elephant and the hippopotamus! The elegance of the camel!

But the butterfly, you will say, a flying flower! I can imagine one as large as a hundred worlds, with wings whose shape, beauty, colours, and motion I cannot even express. But I see it—it flutters from star to star, refreshing them and perfuming them with the light and harmonious breath of its flight! And the people up there gaze at it as it passes in an ecstasy of delight!

What is the matter with me? It is He, the Horla, who haunts me, filling my head with these foolish things! He is in me. He is becoming my soul. I will kill Him.

AUGUST 20. I will kill Him. I have seen Him! Yesterday I sat down at my table and pretended to write very assiduously. I knew quite well that He would come prowling round me, quite close to me, so close that I might perhaps be able to touch Him, to seize Him. And then—then I should be filled with the strength of desperation. I should have my hands, my knees, my chest, my forehead, my teeth to strangle him, to crush him, to bite him, to tear him to pieces. And I watched for him with all my overexcited nerves.

I had lit both my lamps and the eight wax candles on my mantelpiece, as if, by this light I should discover Him.

My bed, my old oak bed with its columns, was opposite me. On my right was the fireplace. On my left the door, which was carefully closed, after I had left it open for some time, in order to attract Him. Behind me was a very high wardrobe with a looking-glass in it, which I used every day to shave and dress by, and in which I always inspected myself from head to foot each time I passed in front of it.

So I pretended to be writing in order to deceive Him, because He also was watching me. And suddenly I felt, I was certain, that He was reading over my shoulder, that He was there, almost touching my ear.

I got up so quickly, with my hands extended, that I almost fell. Horror! It was as bright as midday, but I did not see myself in the glass! It was empty, clear, deep, full of light! But my figure was not reflected in it—and I, I was opposite it! I saw the large, clear glass from top to bottom, and I looked at it with unsteady eyes. I did not dare advance; I did not venture to make a movement; feeling certain, nevertheless, that He was there, but that He would escape me again, He whose imperceptible body had absorbed my reflection.

How frightened I was! A moment later I began to see myself through a mist in the depths of the looking-glass, as if I were looking at it through a veil of water. This water seemed to flow slowly from left to right, making my figure clearer with each passing moment. It was like the end of an eclipse. Whatever hid me did not appear to possess any clearly defined outlines, but was a sort of opaque transparency, which gradually grew clearer.

At last I was able to distinguish myself completely, as I do every day when I look at myself.

I had seen Him! And the horror of it remained with me, and makes me shudder even now.

AUGUST 21. How could I kill Him, since I could not get hold of Him? Poison? But He would see me mix it with the water. Besides, would our poisons have any effect on His impalpable body? No—no, they would not. Then how?—How?

AUGUST 22. I have sent for a blacksmith from Rouen and ordered iron shutters for my room, such as some private hotels in Paris have on the ground floor, for fear of thieves. And he is going to make me a similar door as well. I have made myself out a coward, but I do not care about that!

SEPTEMBER 10. Rouen, Hôtel Continental. It is done. It is done—but is He dead? My mind is thoroughly upset by what I have seen.

Yesterday, the locksmith having put on the iron shutters and door, I left everything open until midnight, although it was getting cold.

Suddenly I felt that He was there, and joy, mad joy took possession of me. I got up softly, and I walked to the right and left for some time, so that He might not guess anything. Then I took off my boots and put on my slippers carelessly. Then I fastened the iron shutters and, going back to the door quickly, I double-locked it with a padlock, putting the key into my pocket.

Suddenly I noticed that He was moving restlessly round me, that in his turn He was frightened and was ordering me to let Him out. I nearly yielded, though I did not quite, but putting my back to the door, I half opened it, just enough to allow me to go out backward, and as I am very tall, my head touched the lintel. I was sure that He had not been able to escape, and I shut Him up quite alone, quite alone. What happiness! I had Him fast. I ran downstairs into the drawing-room which was under my bedroom. I took the two lamps and poured all the oil on to the carpet, the furniture, everywhere. Then I set fire to it and made my escape, after having carefully double locked the door.

I went and hid myself at the bottom of the garden, in a clump of laurel bushes. How long it was! How long it was! Everything was dark, silent, motionless, not a breath of air and not a star, but heavy banks of clouds which one could not see, but which weighed, oh so heavily, on my soul.

I looked at my house and waited. How long it was! I already began to think that the fire had gone out of its own accord, or that He had extinguished it, when one of the lower windows gave way under the violence of the flames, and a long, soft, caressing sheet of red flame mounted up the white wall, and kissed it as high as the roof. The light fell on to the trees, the branches, and the leaves, and a shiver of fear pervaded them also. The birds awoke; a dog began to howl, and it seemed to me as if the day were breaking. Almost immediately two other windows flew into fragments, and I saw that the whole of the lower part of my house was nothing but a terrible furnace. But a cry, a horrible, shrill, heart-rending cry, a woman's cry, sounded through the night, and two garret windows were opened! I had forgotten the servants! I saw the terror-struck faces, and the frantic waving of their arms!

Then, overwhelmed with horror, I ran off to the village, shouting: "Help! Help! Fire! Fire!" Meeting some people who were already coming on to the scene, I went back with them to see.

By this time the house was nothing but a horrible and magnificent funeral pile, a monstrous pyre that lit up the whole country, a pyre where men were burning, and where He was burning also, He, He, my prisoner, that new Being, the new Master, the Horla!

Suddenly the whole roof fell in between the walls, and a volcano of flames darted up to the sky. Through all the windows which opened on to that furnace, I saw the flames darting, and I reflected that He was there, in that kiln, dead.

Dead? Perhaps? His body? Was not his body, which was transparent, indestructible by such means as would kill ours?

If He were not dead? Perhaps time alone has power over that Invisible and Redoubtable Being. Why this transparent, unrecognizable body, this body belonging to a spirit, if it also had to fear ills, infirmities, and premature destruction?

Premature destruction? All human terror springs from that! After man, the Horla. After him who can die every day, at any hour, at any moment, by any accident, comes He who can die only at his appointed day, hour and minute, when He has attained the limits of his existence.

No—no—there is no doubt about it—He is not dead. Then—then—I suppose I must kill myself.

> **Guy de Maupassant** (1850–1893), a French author, was a master of the short story form. "The Horla" is one of his most famous stories, and some critics have suggested that its treatment of insanity was a response to the author's syphilitic headaches, and a conviction that he too was succumbing to a long family history of mental health issues. Although I disagree with those critics, the story was prophetic: five years after its publication the author attempted suicide. He died, insane, at the age of 43.

Commentary

We can interpret "The Horla" as an account of the narrator's descent into madness, a tempting reading because it prefigures Maupassant's own downward spiral into insanity. The author contracted syphilis in his youth and it led to paranoia in his thirties. Four years after publication of "The Horla", he tried cut his own throat. Doctors committed him to an asylum, but he died from his disease a year later. So yes, we can read his story as a tale of madness. But we can also read it as a classic science fiction horror story: the tale of something coming to supplant the human species.

In "The Horla", Maupassant mentions two ways in which the human story might come to an end.

First, the process of evolution might lead to a creature that supplants us.

As the narrator writes at one point: "A new being! Why not? It was assuredly bound to come! Why should we be the last?" This process, indeed, has happened to human species in the past. About 40,000 years ago, modern Homo sapiens supplanted the last of their cousins in Europe, the species of human we call Neanderthals. A different species of human, the Denisovans, once interbred with humans and Neanderthals. The Denisovans appear to have outlasted the Neanderthals, but they died out too. Might not some new species evolve to supplant Homo sapiens?

Well, we can bet that humans will one day become extinct. All species do. But one struggles to imagine other creatures evolving to take our place as the dominant *technologically advanced* species on Earth.

Suppose some catastrophe got rid of humanity. Nuclear war, perhaps, or climate change. Life on Earth would get on just fine in our absence. But no compelling reason exists to suppose the mix of factors contributing to the success of Homo sapiens—a high level of general intelligence; ability to communicate abstract thoughts using a complex grammar; opposable thumbs; and all the rest—would evolve in some other species. Evolution did not have humans in mind when the first life arose. A myriad of contingent events led to us. With humans gone, evolution would work its magic on whatever remained. But why should the process lead once again to intelligent creatures that master technology? I don't believe it would. Similarly, with humans in place, could any creature evolve and then outcompete us on the basis of intelligence? (If evolution worked on the best clay it now possesses, and a creature such as the Horla developed, then we know what would happen: Homo sapiens would end that life form before it became a threat.)

A more plausible scenario involves the products of our technology rather than creatures arising from 'natural' evolution. Perhaps our 'mind children' might supplant us or somehow merge with Homo sapiens. Humans might become posthumans. Science fiction writers have of course explored this concept, with authors considering how genetic engineering, nanotechnology, and artificial intelligence, either singly or in some combination, might offer a route to posthumanity. One can imagine how these technologies could lead to something like the Horla. (By way of contrast, one can also imagine how these technologies could improve everyone's lives and lead to a golden age for humanity. The highlighting of such possibilities constitutes one of the jobs of science fiction.)

Maupassant hints at a second scenario for the demise of humans.

The narrator writes: "There was no moon. The stars glittered in the black depths of the sky. Who inhabits those worlds? What forms, what living beings, what animals dwell there? Do the philosophers on those distant worlds know more than we do? What can they do more than we? What do they see that we do not know of? Will not one of them, some day, cross the gulf of space and appear on our earth as a conqueror, just as formerly the Norsemen crossed the sea to subjugate nations feebler than themselves?"

There we have it. Humans supplanted by extraterrestrials. The classic scenario of science fiction. And Maupassant considered the possibility a decade before H. G. Wells published his novel *War of the Worlds*. But is this scenario plausible? Might alien invaders pose a more dangerous threat than our own technological offspring?

Some scientists believe alien civilisations do indeed present humanity with an existential threat. We saw in Chap. 3 that some scientists object to METI, the sending of messages into the cosmos. They fear malign intelligences might intercept those messages and determine their source. If hostile creatures came here with the intention of wiping us out, humanity would stand no chance.

Any civilisation capable of traversing hundreds of light years to reach us would have little trouble in erasing humanity once it got here.

Other scientists believe extraterrestrial intelligence would have advanced beyond us in ethics as well as technology. We have nothing to fear because alien civilisations would operate under some sort of *Star Trek* Prime Directive. Besides, why would they bother to wage war against us across interstellar distances? (To which the first group of scientists might reply: they might not trust us to operate a Prime Directive ourselves. Humans might not pose a threat to other civilisations *now*, but we might become a threat. In the words of Sergeant Jablonski in *Hill Street Blues*: "let's do it to them before they do it to us". Better to nip a problem in the bud …).

So will some Horla-like creature one day traverse space and appear on our Earth to conquer it? Or will we one day join a peaceful United Federation of Planets, as *Star Trek* would have it? Well, we have no idea of the motivations of extraterrestrial intelligent beings. They might resemble the benevolent Overlords of Arthur Clarke's *Childhood's End*. Or they might display more sinister motives, like the Horla in Maupassant's story.

Another possibility of course exists.

Despite years of listening, we have yet to hear from extraterrestrial beings. If that silence endures then perhaps we must conclude that ours is the only intelligent species in the galaxy. We will then have no need to fear creatures like the Horla. And if we gain the wisdom needed to meet the threats we face here on Earth then—if we survive—we might go on to establish a presence on planets throughout the Galaxy. In that case, nothing could supplant us. But even in that future, humans would change. Humans would change in ways we struggle, from this vantage point, to imagine.

Notes and Further Reading

the author's downward spiral into insanity—For a recent English-language biography of Maupassant, see Lloyd (2020).

human species in the past—In a fascinating book, Reich (2018) describes how the extraction and analysis of ancient DNA from human skeletal remains has led to a scientific revolution. Advances in the technology for extracting DNA has transformed our understanding of human biological evolution, and geneticists are now tracing the relationships between different species of humans—Denisovans, Neanderthals, and Homo sapiens.

our 'mind children' might supplant us—Hans Moravec (1988) introduced the concept of our 'mind children', superintelligent thinking machines that are our 'artificial progeny'. Such machines, operating at levels beyond our understanding, might pose a threat to humanity.

a decade before H. G. Wells—Wells wrote The War of the Worlds between 1895 and 1897, and it appeared as a serial in magazine form in 1897. It first appeared in book form in 1898. Maupassant published a short version of "The Horla" in 1886 and a longer version, the form reproduced here, in 1887. So it is fair to say that the Maupassant short story appeared a decade before the Wells novel.

References

Lloyd, C.: Guy de Maupassant. Reaktion Books, London (2020)
Moravec, H.: Mind Children: The Future of Robot and Human Intelligence. Harvard University Press, Cambridge, MA (1988)
Reich, D.: Who We Are and How We Got Here: Ancient DNA and the New Science of the Human Past. Oxford University Press, Oxford (2018)
Wells, H.G.: The War of the Worlds. Heinemann, London (1898)

GPSR Compliance

The European Union's (EU) General Product Safety Regulation (GPSR) is a set of rules that requires consumer products to be safe and our obligations to ensure this.

If you have any concerns about our products, you can contact us on ProductSafety@springernature.com

In case Publisher is established outside the EU, the EU authorized representative is:

Springer Nature Customer Service Center GmbH
Europaplatz 3
69115 Heidelberg, Germany

Batch number: 08050695

Printed by Printforce, the Netherlands